POSTGRADUATE STUDY IN THE BIOLOGICAL SCIENCES

A Researcher's Companion
by
Robert J. Beynon

Published by Portland Press Ltd., 59 Portland Place, London
W1N 3AJ, U.K.
In North America orders should be sent to Portland Press Inc.,
Ashgate Publishing Co., Old Post Road, Brookfield,
VT05036-9704, U.S.A.

© **1993 Portland Press, London**

ISBN 1 85578 009 7

British Library Cataloguing in Publication Data
A catalogue record for this book is available from the British Library

Designed by Angie Moyes
Typeset by Portland Press, Ltd
Printed in Great Britain by Cambridge University Press

C Contents

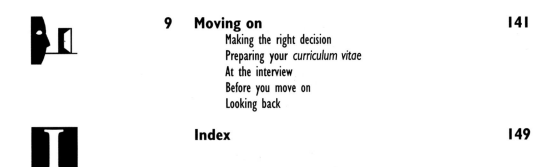

Preface

This guidebook has developed from a brief postgraduate guide that I prepared for research students in the Department of Biochemistry at the University of Liverpool. In producing this book, I feel very exposed. I offer advice that I do not always follow myself. I advocate clear writing; this book will be scrutinized for the clarity of my text. I advocate good design and clear graphics, the use of word processors and spelling checkers — you, the reader, are therefore entitled to castigate me if I fail on any of those counts.

To supervisors: what entitles me to write this book? Frankly, nothing, other than the role I have played as student and as supervisor. I am concerned that so many students are thrown in 'at the deep end' of a research programme that they quickly become immersed in the science, and never again have much time for other activities that make a professional scientist. They are expected to know so much, so suddenly, that a few hints and ideas may be welcomed. My aim is always to set the student thinking, to show them a few examples, give them a few ideas, but persuade them to discover their own solutions.

To students: there is more to science, and to a research project, than laboratory benchwork. I do not intend this book to be read as dogma. If I appear to make bold statements, treat them as suggestions and develop your own ideas, but please think about the subject matter herein. I am sure that analysis of my text would reveal a recurrent theme: **plan**. The thesaurus on my computer gives me almost two hundred words that associate with 'plan', including: sketch, draft, schedule, budget, allocate...One of the most important skills that you should seek to acquire is to be able to 'plan' in most of these 200 subtly different ways!

Acknowledgments

I am grateful to Portland Press for letting me do this book my way, and for providing total support at all times. In particular, I am grateful to Rhonda Oliver for careful editing and advice, and for her enthusiasm for this project. I thought that someone should act as a referee, and I am grateful to John Lagnado for many valuable suggestions.

Undoubtedly, my own students, past and present, will come across this book and observe wryly that this is not the way things were or are, even in my own laboratory. It is their successes and their difficulties, however, that persuaded me of the need for this book, and it is dedicated to them.

Robert J. Beynon
1993

Starting out

Making the right decision

Let us assume for the moment that you have yet to make the decision to enter a research career by becoming a postgraduate student. What do you want from your higher degree? How much of yourself are you prepared to invest to achieve these aims? Let us first identify, and then dismiss, some negatives: relatively poor financial reward, long working hours, the periods of desolation when nothing seems to work, the struggle of assimilating endless new information, and the stressful efforts of learning to communicate your science.

Now, the benefits. You will acquire, in no particular order of importance: a detailed knowledge of one small area of your subject, considerable skills in experimentation, a higher qualification, a thesis and, usually, papers published in the scientific literature. You may acquire some teaching experience. You will probably attend scientific meetings in your country, and perhaps in foreign countries, and will have access to the worldwide, largely apolitical scientific community and the results of their research. You may make some lifelong friends during this time. You will generally increase your chances of permanent employment as a professional scientist. You may make a valuable contribution to the sum total of human knowledge. But mostly, you should enjoy what you are doing, even in the bleak periods.

If you decide to obtain a higher degree by research, you will have to make a series of value judgments before applying to your chosen laboratories. What subject, which institution, which supervisor? There is no way that this book can guarantee that you will make the correct choices. Take heart that nearly all decisions will turn out to be right — most students complete their studies successfully. However, asking the right questions before you start can help, by matching you as closely as possible to a supervisor and laboratory that will enable you to make the best of yourself.

You, the student

What qualities are sought in a research student? Ideally, you will be totally dedicated, tireless, imaginative, an intellectual giant and a technical wizard, possessed with an overwhelming humility. In reality, you will probably be like most other scientists: a normal human being. If you worry about being clever enough, consider what Peter Medawar had to say in 'Advice to a Young Scientist':

"One does not need to be terrifically brainy to be a good scientist...Common sense one cannot do without, and one would be the better for owning some of the old-fashioned virtues...application, diligence, a sense of purpose, the power to concentrate, to persevere and not to be cast down by adversity..."

Medawar's approval of these 'old-fashioned virtues' recognizes that scientific research is a rigorous, demanding and sometimes lonely occupation. At some stage in their studies, most graduate students probably consider giving it all up. This is usually a side-effect of a period of frustration that evaporates once the student has brought their work back on track. For the rewards, try Carl Sindermann in '*The Joy of Science*':

"...some of the 'joys' are those of insight and discovery, of producing proof through experimentation and observation, of interactions with peers, and even of communication with the public. Among the joys are a sense of quiet satisfaction and well-being, and a feeling of internal security — knowing that it is possible to function effectively as a professional among professionals"

The subject

Do not be too restrictive about the subject matter of your research. By all means, identify broad areas in which you would like to work or techniques that you would like to acquire as a speciality, but remember that your research programme is primarily a training in becoming a scientist, and that you should be able to acquire this training in almost any subject area. Of those graduate students who go on to conduct postdoctoral research, many will change their subject matter. They will move into this new area armed with the hard-won skills of scientific research management, of research method, of diligent experimentation, and of communication. To their new subject area they apply these skills with ease. Some will seek and welcome this 'fresh start' as an opportunity to apply their skills. Others will wish to maintain the subject matter of their original research programme, making best use of specific skills to launch the next stage of their careers.

The laboratory

If at all possible, visit the laboratory before you accept an offer of a research studentship. There are a number of questions that you should seek to answer. Do you have any preconceptions of what a well-run laboratory should look like? Whether cluttered or immaculate, neither appearance gives much indication as to how well the laboratory works. How many others work in the laboratory? Are you the sort of person who would prefer to work alone, with your own research topic? Or, could you survive in the hustle and bustle of a large group, where you may be working on a small part of a problem that is shared with several others? How well-equipped is the laboratory? How much bench space will you be given? Does the laboratory give the impression of being industrious and efficient, and the people working in it friendly and enthusiastic? No one of these issues is of over-riding importance, and many are mutually exclusive. You cannot expect a lot of personal space in a packed laboratory, but you will have access to the experience, skills and support of a large group working in the same area. Try to talk to other graduate students in the laboratory. Ask them about the good points, but also try to discover any bad points of working there.

The department and institution

Try to find out more about the department, and in particular, what it considers to be its responsibilities towards graduate students. Increasing pressure is being brought to bear on departments to formalize some aspects of postgraduate training but the response has not been uniform in its enthusiasm. Go armed with these questions:

— is there any specific coursework for graduate students?

— what will you need to produce as progress reports and presentations?

— how many graduate students are there?

— how many students submit their theses on time?

Within reason, the more you are asked to do, the more likely it is that the department has a good attitude to graduate student training.

At the heart of your research programme will be your relationship with your supervisor, and your choice of supervisor is pivotal. Within reason, a good supervisor can be found in almost any department in any institution. Avoid the temptation to use labels that mark institutions or departments as 'good' or 'bad' in research; such yardsticks can reflect the number as well as the quality of practitioners, and good and bad scientists or supervisors are to be found in all institutions.

Your supervisor

More will be said in the next section about the relationship that you will develop with your supervisor, but first you have to choose each other. This usually entails a tour of the laboratory and department, and an interview in which to make up your minds. You will have prepared a *curriculum vitae*. The person sitting opposite you will be most interested in your research potential, and is more likely to talk about your undergraduate research project than details of your coursework. If you are enthusiastic about working in this laboratory, you will have done a little background reading on the supervisor's stated research interests, or looked up his or her name in a citation index or an abstracting service to discover what they have published in recent years. If you make sure that you impart this background research you will make a good impression. Be sure to ask questions about the research project. Probably, the line should be drawn at requesting the supervisor's *curriculum vitae*! Ideally, you will want a good scientist who is also a good supervisor; you will probably survive with a good scientist who is not such a good supervisor, but you will fail to make the best of yourself if you choose a mediocre scientist, irrespective of their abilities as a supervisor. At the end of the day, ask yourself whether you could work with (not 'for') this individual for the next three or more years.

The student/supervisor relationship

During your research, you will develop a close relationship with your supervisor, about whom you will have a gamut of opinions ranging from respect and awe to irritation and contempt. Sometimes you will run through all of these emotions in a single day!

Overheard in a laboratory:

"What is the difference between a god and my supervisor? A god is here but everwhere, my supervisor is everwhere but here."

Individual supervisors differ considerably in their interpretation of their role and responsibility to their students. Some adopt a 'sink-or-swim' approach and leave the student to their own devices; at the other extreme, some will 'mollycoddle' their students, do most of the intellectual work for them and may even, in effect, write their theses. Most adopt a sensible intermediate between these two extremes. Some like formal meetings at regular intervals; some like casual chats over your laboratory bench. Others like regular group meetings. The extent of contact between you and your supervisor will vary as the project progresses and develops, but it is wrong to assume that the responsibility for the frequency and duration of discussions is solely your supervisor's. It is your research programme, and if you feel the need for a long or short discussion with your supervisor, it is up to you to make sure that it takes place. Request that your supervisor set aside some uninterrupted time, and a relatively clear desk, so that you can work through your data and ideas. Be prepared for these meetings; have a clear idea of what you want to discuss, and give some thought to the next stage of your research. This is part of claiming your research project as your own.

Competition for your supervisor's time will be fierce, with administration and teaching, other external responsibilities and other research commitments all plundering his or her day. Enjoy the luxury of working in a well-equipped and well-funded laboratory with six others, but do not expect the undivided attention of the person who found the grant support for all this! The prime time for demands on your supervisor's is 9 am to 5 pm, Monday to Friday. Why not suggest a discussion meeting outside these hours, perhaps early in the morning, early evening or at a weekend? You will benefit, because the chances of the meeting taking place without interruptions are so much higher and your supervisor will benefit from being able to give you undivided attention, and from the pleasure of seeing the student make an effort to assert their responsibility for the project.

Critic or coach?

At times, the behaviour of your supervisor will be difficult to understand. If a series of experiments is going well, your mentor might seem rather negative about the whole thing, asking for more replicates or controls, or seeking alternative explanations. It may not seem so at the time, but this is precisely because your supervisor is as intrigued by the data as you, but is showing, indirectly, one of the qualities of a good scientist — self-criticism. Conversely, when your work seems to be in the doldrums, and you are making little progress, you might find that your supervisor seems excessively optimistic, seeking out the positive aspects of your work, and suggesting other solutions. Now you are being shown how to hone your skills when they are most needed, at an *impasse*. There will be times when you are as helpless as each other. After three years of playing this game of *alter ego*, it would be surprising if you had not had some moments of disagreement, as well as periods of shared excitement and elation.

Simple differences of opinion are usually forgotten immediately, and the outcome is a better understanding of each other. Very rarely, disagreements between you and your supervisor may escalate to the extent that it is not really possible for you to continue to work together. Your options at this stage depend upon your department. Possibly you could transfer to another supervisor, but it is likely that this will entail starting a new project. If the studentship was awarded to your supervisor, rather than to the department, you would not be able to retain the same stipend. Before doing anything, talk about the problem to someone (another member of staff or the Head of Department) who would be sympathetic and who could act as an intermediary. Consider whether the best course would be to discuss your feelings directly with your supervisor, perhaps with another person present. Problems may evaporate during this discussion, and you may find a new respect for each other. In turn, that respect allows you to seek a pattern of working together that is suited to you both.

Wrong decision?

The 'drop-out' rate for postgraduate students in the biological sciences is low, but there are some who discover, usually towards the end of their first year, that they are not suited to a research career. You have not 'failed'. The system rarely provides you with any prior experience of real research, and it is unfortunate that some will discover that they are unsuited to research when they have entered the course. If you feel uneasy about going on, speak to your supervisor as a matter of some urgency. Together, you should be able to work out the best way to allow you to re-route your career.

Rules and regulations

As a graduate student, you will be bound by regulations or ordinances of the institution that will award you your higher degree. Additionally, the department may impose further requirements upon you, such as the obligation to attend seminars, present a departmental lecture, and produce a poster communication. The funding body that provides your maintenance, fees and running costs may also have specific requirements. These requirements will be so specific to your local department and institution that it is not possible to generalize, other than to offer one piece of advice. Do not expect your supervisor to provide the impetus when you write your first-year report, or prepare a seminar. Nor must you expect your supervisor to set a date on which you should start to consider these activities. It is **your** research degree, and your responsibility to find out what is expected of you, and to meet those expectations. Even better, try to exceed them. Offer an outline of your report, or a sketch of your poster before it is requested. Show that you can be as professional in the management of the formalities of your research degree as you are in the prosecution of your research.

See Chapter 3, 'Conducting a research project'

Find out, well in advance, what obligations you have to meet. Read the regulations, and extract important items. What time scales apply to you? What specific guidance is given about your thesis? Must any notices of intention to submit a thesis precede submission by a specific time period. Work back through the regulations, from the *viva voce* examination to the start, and mark off important times and events.

Working with others

A research laboratory is often a crowded environment, filled with individuals who have been selected on the basis of their independence of thought, strength of conviction and general self-confidence. These individuals are competing for space, equipment, shared facilities and their supervisor's time. Given such a potentially explosive situation, it is a tribute to them that for most of their time they work together in harmony. Yet, occasionally, tensions do develop, often over trivial matters. A communal solution used up, but not replaced; music in the laboratory, or equipment left uncleaned can all precipitate a difference of opinion that spreads through the whole laboratory, increasing tension. These differences are never serious, and should be dispelled as quickly as possible. If you can resolve the issue in the laboratory without involving the supervisor, so much the better, but if that fails, you should not be wary of calling such differences to attention. Sort it out, and sort it out quickly. Neither party really wants to be involved in a cold war, nor do the rest of the laboratory.

Many misunderstandings arise because a newcomer works in a way that differs from accepted laboratory lore. As a newcomer to the laboratory, be cautious that there will be an unwritten 'rule book' of the 'right' way to behave. Try to be sensitive to this, and aim to discover the 'rules' as quickly as you can, by asking others in the group. In turn, when you have been in the laboratory for a year or so, spare a thought for those who succeed you, and help them to fit in as well.

Further reading

Medawar, P. B.(1979) 'Advice to a Young Scientist', Harper & Row, New York, London

A book that, unfortunately, is most likely to be read at a later rather than earlier stage in any scientific career. Try to read it at your earliest opportunity. Delightfully written, and full of good advice and thought provoking ideas; everyone should find a part that strikes a chord.

Sindermann, C. J. (1985) 'The Joy of Science', Plenum Press, London

An extremely readable and sometimes provocative book that is most germane to scientists from the U.S.A. at the postdoctoral level and above. Nonetheless, it offers good advice, is enjoyable to read, and may provide some insights into the modus operandi *of your supervisor and other colleagues.*

Phillips, E.M. & Pugh, D.S. (1987) 'How to Get a Ph.D.', Open University Press, Milton Keynes

Perhaps containing more lessons for supervisors than students, this book nonetheless identifies many of the crises that can occur, and offers good advice on dealing with them.

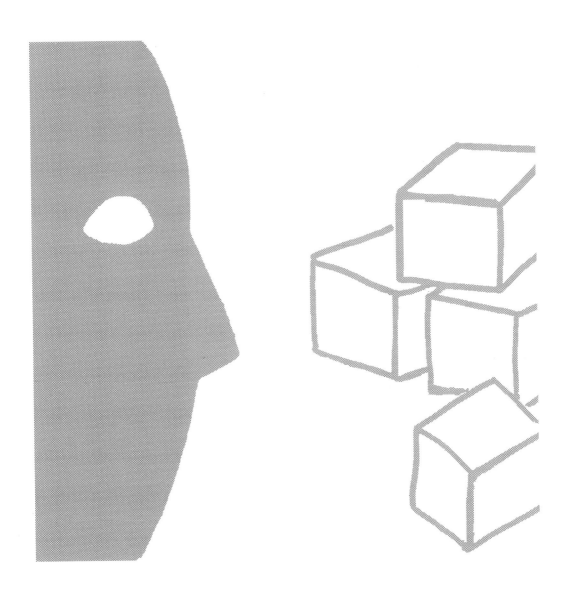

Developing your
experimental skills

Introduction

Make a quick mental survey of the postgraduate students in your department and note the diversity of 'projects' that they conduct, all of which will lead to the award of a higher degree. Some projects will emphasize the acquisition of data, with analysis being a late event; others will develop theories that are subsequently followed up by experiment. Still others will devise new methodologies that will be used more by their successors than by themselves. Some students will spend most of their time in front of a computer terminal, others seem to be permanently poring over a microscope, spectrophotometer, h.p.l.c., electrophoresis apparatus, mass spectrometer or any of the other pieces of equipment that are the tools of modern science. Some will be found in out-patient departments waiting for a new blood sample. There are projects that will require the student to be away from base for extended periods, perhaps in another country.

This diversity of research, and indeed, the diversity of personalities of the people conducting them, complicates and enriches the definition of a research project. What is it that binds all of these types of activities into a common structure? For what exactly will you be given a higher degree? A common standard answer to this question is that you will have been trained in, and have demonstrated that you can apply, the processes of 'the scientific method'. Yet, nowhere in your training will you have been given explicit instruction in this skill.

Learning the scientific method

It is rare to find classes that specifically address this issue of 'scientific method', and this is unlikely to be a subject that you were taught in your undergraduate days. At best, you will have conducted a research project in your final undergraduate year; indeed, it is probably this experience that tempted you to a higher degree. Yet at the end of your research programme, you should be in a position to conduct research as an independent investigator and even to supervise other incoming research students. How do you acquire these skills?

Certainly, the scientific literature gives no clue as to the way in which science is done; nearly all scientific papers are written as though the science was conducted in a logical progression (Medawar goes so far as to suggest that scientific papers are a 'fraud'). Science as reported in journals is far removed from the everyday science that is conducted in laboratories. Papers contain little or no acknowledgment of the chronological development of the data and the theories to explain that data, and no paper ever reports the spilt enzyme sample, the contaminated DNA, or the fact that a month's incomprehensible data was eventually explained when a new control experiment was devised. No paper shows the false starts, or the sudden change in hypothesis necessitated by a new experiment.

Reading books about the nature of experimental science will not give you the set of intellectual skills that define a practising scientist. Few scientists apply a strictly formal method of scientific reasoning that is found in such books. The way in which you approach scientific puzzles will be conditioned as much by your personality as by your training, and it is difficult to recognize and transfer the uniquely personal skills that make a successful scientist. These are acquired by trial and error, with a great deal of disappointment, and periods of extreme frustration. You cannot be taught to be the same type of researcher as your supervisor; their skills are the results of their personal experience and development; yours will be equally unique. At best, you can be shown how you should think and plan, and from these beginnings you will shape your own research potential. Can we define those skills that characterize a good scientist? They might include:

- a good background knowledge, coupled with an understanding of where this knowledge is most certain or most tenuous

- the ability to shape hypotheses or theories, to conceptualize models of unseen processes and to let these models suggest experiments

- a finely honed ability to design a properly controlled experiment, in which all factors have been taken into account

- an understanding of data gathering and recording that extends to the limitations and strengths of the tools that generate the data

- the ability to see all of the data, rather than that which is most prominent or desirable.

All of these qualities are obvious, but, sadly, many of them are never given a moment's thought during postgraduate research. You will undoubtedly acquire these skills, to a greater or lesser extent, but will you ever consider them in an abstract sense? When the opportunity presents itself, try to stand back from your day-to-day activities and think about your research from perspectives other than those dictated by the need to get the next experiment done.

Experimental design

Experimental design is so dependent on the subject matter that there are few rules that are generally applicable. You will discover how experiments are done in your chosen field, and from your reading of the literature will eventually be able to discriminate between a 'good' and a 'bad' experiment. In this section are given a few of the most obvious, and most important, aspects of good experimental design.

Identifying the variables

The identification of variables is so fundamental a skill that you may not even be consciously aware of what you do when you incorporate this thinking into experimental design. You will be aware that the dependent variable (the one that you are measuring) will be affected by the independent variable (the one that you vary according to your needs). If you have two independent variables (for example, time of incubation and sample concentration), you should realize that you are conducting two experiments simultaneously. Make sure that each of these experiments can be analysed independently of the other, and that the appropriate permutations and combinations of the two independent variables have been included.

Be wary of 'hidden' uncontrolled variables that can creep into your experiments. Addition of an unbuffered reagent may cause a pH shift in the reaction; residual salt in a nucleic acid sample can have a significant influence on the ionic strength of the reaction mixture; temperature changes modify other parameters such as pH. Learn how each of the parameters interacts, and establish that each of them has been properly controlled, even if at first glance it appears that they are not the subject of the experiment.

Controlled experiments

A controlled experiment is one in which any of the outcomes can be interpreted in terms of the hypothesis that led to the original design. This assumes that you manipulate the independent variables within the system, such that the only parameters that change are under your control.

Negative controls

A negative control is a precaution against 'false positives'; it ensures that no other component in the system could give rise to the effect that you are attributing to the key component. Omit that key component in the experiments to ensure the lack of a demonstrable effect. Such experiments are thus protected from misinterpretation due to sample cross-contamination, spurious effects from other sources and un-expected interactions between components of the experimental system.

Some negative controls include all the components that are present in the test experiment, but apply them in a different order, or in such a fashion that the effect would be prevented. For example, add the enzyme inactivator at the start rather than the end of the incubation, or use heat-denatured enzyme.

Positive controls

Positive controls serve a different purpose. There are two types of positive controls: those that test the whole experimental system *per se*, and those that test that the system is functioning correctly each time it is used. Positive controls of the first type can be difficult, and you might have to invest considerable effort to generate a control system that will prove your experimental system. They are tests of the experimental model themselves, and are designed to include material that you know will generate a good result. They also provide a good test system to explore the limitations of the experimental system.

The second type of positive control is included every time the experimental system is used. If an experiment is controlled in this fashion, you will always be able to state with certainty that a negative result was genuinely due to a lack of effect of your test material, rather than a one-off failure of the experimental model (for example, a reagent that has 'gone off', or the omission of a key component).

Remember that the positive and negative controls are needed not only in the 'main' experiment but also to control subsequent analytical techniques; e.g. the 'standard' and 'blank' samples. Consider the full range of results that you might get, and ask yourself whether you can explain any such result.

A 'walk through'   When you design experiments, you have two aspects to consider: the logic underpinning the experiment and the practical implementation. The logic includes the identification of the variables and the inclusion of positive and negative controls. Re-examine whether the experiment is a true test of the hypothesis that underpins your thinking. Consider the data that you will derive. Do you have any expectations or prior knowledge of the range of values that you might expect? If these can be calculated, even if you have to use simplifying assumptions, it will help you decide on the practical implementation.

A walk-through should also be used to consider the practical aspects of the experiment. The considerations are now different, and are dictated by such factors as availability of starting material, the amount of sample that might be needed for the final analysis, the volumes that you might need for the analysis and the need for replicated analyses. A quick sketch of the experimental design will allow you to plan the scale of the experiment, and calculate the volumes and quantities of the reagents that you need; an experiment in which you are forced to make up fresh reagent half way through immediately loses one of its tight controls. If possible, plan experiments in such a way that you will not analyse all of the material in the last step; reserve the possibility for re-analysis of an otherwise successful experiment.

Many experiments include time as the independent variable, and you will have to design time intervals. A natural tendency is to take time points at equal intervals, but this is rarely the best choice. If you do not know the time course of a process, use intervals that start small and which become progressively larger (e.g. 1, 2, 4, 8, 16, 30 min, 1 h, 2 h). This will improve the chances of collecting useable data whether the process is rapid or slow. This is an example of a ranging experiment that has the goal of defining the time course of the process. Similar ranging experiments can be conducted to define sample size, linearity of analyses and so forth.

Aim of an experiment   Always define the purpose of the experiment before you embark upon it; an ill-planned experiment is more likely to fail. Experiments should not be made to serve too many purposes at once, and it is quite acceptable for you to devise experiments that have the sole purpose of teaching yourself a new method. Do not be pressurized into trying out a new and unfamiliar technique or method with your carefully won experimental samples. Resist the temptation to do this; experiments involving new methods often fail because it is only at this time that you identify all of the problems and subtleties of the methodology, as well as ambiguities in the published procedure. If possible, try out new methods with samples that are readily available and known to give a suitable response. Indeed, if the pressure of using your own material is removed, you can concentrate much more on the methodology, with a correspondingly greater chance of success, both during the dry run and when the method is ultimately applied to your own material.

Scientific equipment: 'black boxes'?

Most of the data that you derive during your project will be yielded up to you from highly sophisticated equipment. Moreover, analogue output (meters, chart recorders) is being replaced by digital storage, manipulation and output of data in a processed form. Thus, you may not even see the raw data that is generated by the machine. The whole of your work is dependent upon the correct operation of this equipment and of the data-processing algorithms. How much time and effort should you devote to understanding this equipment, and the way in which it manipulates your data? If an item of equipment is being used a great deal, you will be expected to build up your understanding to the level of an expert. There are many things that you can do:

- read the manual (it is amazing how rarely this is done)

- understand any dangers and safety responsibilities that the equipment might need

- have someone take you through the operation of the equipment as often as you feel you need

- test the equipment with proven samples

- 'play' with good data and discover the extent of the data-processing capabilities of the equipment

- learn rudimentary user-level troubleshooting and servicing skills

- develop the discipline of the 'pre-flight' check of the equipment before you use it

See Chapter 6, 'Computers and computing'

- talk to service engineers and technicians when they overhaul or repair equipment, and take the opportunity to examine the innards of the machine

- if the machine is computer controlled, accept that you cannot ignore this aspect of the use of the equipment

- report a breakdown immediately, even if you suspect that you were responsible!

Troubleshooting

A recurrent feature of your postgraduate training will be the experiment that does not work. You can repeat the whole experiment in the hope that it will work properly this time. Simple errors in the practical side of the experimental preparation are usually eliminated in the repetition. What if the experiment fails the second time? A third repetition is not likely to help, and at this stage you must bring all of your analytical powers into action.

A well-designed experiment will contain negative and positive controls that allow you to focus on one aspect as the likely source of the problem. Having done this, you should start from the premise that all aspects of the experimental design are potential sources of failure, and seek to eliminate them, one by one, from your 'suspect file'. If the experiment can be broken down into discrete stages, test each one individually, using standard materials, and using the same solutions and materials as the experiment that failed. Recalculate the progress of material through the experiment to make sure that the final result expected is within the range detectable by your analytical procedures. Finally, expect the unexpected, and be prepared to learn; a 'failed' experiment can be the first indication of a new and different nuance to the behaviour of your system. This is one of the ways in which whole new research avenues are opened.

Statistics

Almost all fields of research will require that you use statistical methods to indicate the certainty with which you may make assertive statements. You would rightly be cautious if you had obtained an experimental result once, and would want to replicate that experiment — but how often? Two sets of data might give different average values, but what is the probability that those averages could be arrived at by random sampling of a homogeneous population? What is the best line of fit through a calibration graph, and what is the certainty that you can attach to values obtained from this graph by interpolation? It is difficult, if not impossible, for you ignore the role of statistical analysis in your work. Your department or institution may put on short courses on statistics, and you will need to consult one of the many statistics textbooks that are available. Most of the statistical tests that you need will be available in a computer-based statistics package, and thus you will not need to resort to manual calculations, notwithstanding the educational value of this. Some of the pitfalls that you must consider are:

- using old-fashioned methods of data analysis, because they are already in use in the laboratory

- making assumptions about the error structure of your data — the most common is the assumption of normality

- leaving the decision about statistical methods until after the data have been collected — consideration of statistics is a component of experimental design

- applying tests that you do not understand

- using statistics packages without really understanding what they are doing with your data

- hiding behind a weakly confirmatory statistical analysis when you know, in your heart, that you should repeat the experiment!

The human factor

An important factor in scientific experimentation is the scientist. Do you always have to be coldly objective about your work? Objective, yes, but cold, no. It is the human qualities that produce the new theories, the flashes of insight, the ability to persevere when things are difficult. Finding the right balance between you, the scientist, and you, the person, is difficult, and you should be aware of the common traps that await you.

Resistance to new ideas

One of the paradoxes of the human aspect of science is the conflict introduced by a new idea. The bolder it is, the more we are enchanted and intrigued, and yet the more we resist, precisely because that is the nature of a scientific training. When is the right time to adopt the new idea? The difficulty is nicely put by Trotter (quoted in Beveridge, 1968):

"The mind likes a strange idea as little as the body likes a strange protein, and resists it with similar energy. It would not perhaps be too fanciful to say that a new idea is the most quickly acting antigen known to science. If we watch ourselves honestly we shall often find that we have begun to argue against a new idea even before it has been completely stated."

Unfortunately, there is no answer to this dilemma; yes, you must question new ideas but, you must not resist them unreasonably. At best, you can be aware of the dilemma and find your own balance.

Adherence to disproved ideas

Another danger, equal to that of resisting a new idea is the tendency to adhere to theories that have been disproved. Some of you will have to accept that the original premises upon which your project was based did not hold out. This of course is a decision for both you and your supervisor. The time may come for the theory to be consigned to the waste bin, and a new theory constructed. It is difficult to know when to let go of your hypothesis, but you must accept that this is the nature of science, and is the basis of the development of scientific thinking. Even more important than letting go of an old idea is the formulation of a new theory that can be tested — this is the real challenge.

Zinsser refers to people who hang on to old ideas as "hens sitting on boiled eggs" (quoted in Beveridge, 1968)

A thesis that disproves an idea is perfectly acceptable, as long as the work has been conducted with flawless logic and rigorous experimentation. Indeed, it is often necessary to work harder to reject something than it is to accept it (whether a hypothesis, a scientific paper or a grant application). Your critical skills will be severely tested. Moreover, in your written thesis, you must also show your skill at creating a hypothesis or conceptual model to accommodate your new data.

The dangers of generalization

You will have come across those individuals who let their personal experiences stand as a general model for the whole of the population, usually in conversation, but sometimes in writing also: 'This happened to me, so it must be true for everyone'. When exposed to such statements in the popular media, you will reject them as unscientific, but be cautious of falling into the same trap in science. It is not uncommon to hear a colleague say 'such-and-such a method doesn't work' and, upon closer questioning, discover that the method was tried once, and there was no attempt to discover why it failed. Another generalization is to extrapolate from a specific example to all related experiments ('I didn't control for X because I did those controls last week'). Beware of these generalizations, both as recipient and originator.

Intuition and serendipity

You will undoubtedly be aware of the old adage that 'chance favours a prepared mind'. To be effective as a scientist, you must develop the ability to look beyond the obvious, and to seek for other reasons and meanings for your data. Often, none will be found, but even if the opportunity happens but once in your career you must be able to seize it. It therefore behoves you to sharpen your analytical skills on any problem that you face, but at the same time, avoid the tendency towards a 'tunnel vision' that will prevent you from seeing the less obvious explanations.

Many of you will present your supervisors with a problem, and after a few minutes of perusal they will offer a suggestion that, while not immediately obvious to you, turns out to be the correct suggestion. This is no mystic skill; it is the product of a trained mind that has amassed a wealth of experiences upon which to draw. Your goal should be to develop your intellect to the same level.

The value of discussion

One of the most rewarding experiences in science is discussion of ideas with interested colleagues. Such discussions are particularly important if you find yourself in a hiatus, and are not really sure how to proceed. Discussion allows you to draw on the experiences of others and their different ways of thinking. The 'obvious' questions that they will ask may challenge you to think your problem back to basics, and re-examine your own assumptions. They may have access to different methodology that could be applied to your own problem.

See Chapter 5, 'Talking about your work'

There are some ground rules for such discussions. You must be prepared to admit your lack of knowledge, and to seek clarification of important points. Aggressive questioning and criticism achieves little other than to silence your colleagues for fear of similar treatment. Be prepared to let go of favourite ideas, and to embrace new ones. Open enthusiasm is sometimes scoffed at, but it is a major catalyst to further ideas, and is infectious — enjoy it!

Be prepared to seek advice from other scientists at the same level as your supervisor. At the beginning of your project, there is a tendency to see your supervisor as a source of all information, and to believe that this information is invariably correct. Soon, you will learn that this is not the case, and that supervisors are also limited in what they know. Indeed, it might be overwhelming to work with a supervisor who did know everything! At the same time, you will hopefully see your supervisor in discussion with, and offering advice to, other students outside your laboratory. The same opportunity is open to you; you need no-one's permission to go and talk to another postdoctoral fellow or member of academic staff, although it is courteous to keep your supervisor in the picture.

Scientific integrity

In recent years, it has become hard to escape the publicity afforded to scientific fraud, usually the deliberate falsification of data. You will of course condemn such practices, but where does fraud begin? A northern blot is cropped for publication to eliminate the additional band of hybridization at a size that is too low to be biologically significant. You produce a calibration graph and one point lies far off the middle of an otherwise excellent calibration line. Do you omit that point when you draw the calibration line? By all means use a statistical method that will allow you to eliminate an 'outlier', but realize that the net effect is the same; you have 'falsified' your data. There are many arguments that you might bring into play to justify this decision, and even this simple example is more complex than at first glance. Whatever the outcome, the two golden rules are to record and retain **all** primary data, and to indicate clearly where you have applied a value judgment to the analysis. If you are ever in doubt, 'fail-safe' and show all of the data. You should not expect to make these decisions alone — discuss the situation with your supervisor.

Further reading

Beveridge, W.I.B. (1968) 'The Art of Scientific Investigation', Heinemann, London

This is an inspiring book that starts from the premise that "...the most important instrument in research is the mind of man". It is full of lessons for all of us, illustrated with examples from many great names in biology and medicine.

Medawar, P.B. (1964) 'Is the scientific paper a fraud?' In' Experiment'
(Edge, D., ed), BBC Publications, London

An interesting paper that discusses the relationship between science as it is
done and as it is reported.

Conducting a research project

Managing your time

When you begin your graduate studies, it might seem that the three years of research ahead of you is an eternity, and that you will have more than enough time to complete a detailed research project. Your thesis would probably be overburdened if you conducted one successful experiment a week. In fact, nearly all graduate students will perform many more experiments than that. Some experiments will fail through carelessness in design or execution, others through the intervention of uncontrolled variables, and others will just not be good enough. All experimental work must be replicated n times, although the value of n is very context dependent. Some experiments, with hindsight, should not have been done, and the time wasted could have been prevented by judicious planning and reading. Blind alleys will be followed, and sometimes external events will precipitate a dramatic change in direction of your project.

Although you will not need to conduct what used to be called a 'time and efficiency' analysis of your studies, there is much to be gained from an appreciation of the need for you to manage your time. This includes long-term management of the whole duration of your studies, and short-term organization of your activities on a weekly or even daily basis. Your supervisor might advise you on these matters, but is unlikely to lay down strict requirements. Part of the process of becoming an effective researcher is in the organization of your resources; one of the most precious of these is your time.

Long-term time management

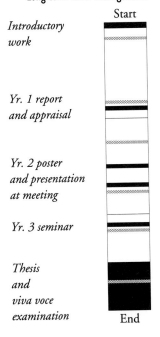

Start

Introductory work

Yr. 1 report and appraisal

Yr. 2 poster and presentation at meeting

Yr. 3 seminar

Thesis and viva voce examination

End

A typical three-year graduate research programme. The white areas are for actual research!

Draw out a long thin rectangle on a piece of paper, and mark it into three equally-sized lengths, as in the example adjacent. Each section will represent one year of a typical three-year research project, although for shorter or longer periods of research, the principle is the same. The whole rectangle represents your research programme, including the time required to write and submit your thesis. For thesis production, you should mark out a block that occupies about one-half of your final year. Now add in any time that you will be spending away from your home base, such as in a collaborator's laboratory. Add further blocks for the various responsibilities that you will have to undertake during your programme of study, such as a first-year report, a poster presentation and a departmental seminar — the example adjacent is fairly typical, and is based on the author's institution. Assess realistically how long it will take you to prepare these items (typically two weeks for each). Realize that you can mark off about a further four weeks each year for vacations (shown in grey). You have just managed to reduce your research time from three to two years (the residual white space)! In that remaining time you must become proficient in the design, execution and analysis of a series of experiments that form a coherent and logical development of your ideas.

You need to consider the allocation and efficient use of this time. Discuss with your supervisor the periods that you might be expected to invest in acquiring the basic methodology, in initial explorations or in peripheral (but important) work. Ask whether warning flags should be raised if you have not reached a particular stage on schedule. Discover whether there are contingency plans that can be brought into play should it transpire that the original concept of your thesis is unworkable (through nobody's fault; this is the nature of experimental science). What arrangements will you have for regular meetings with your supervisor in which you both review your progress, rather than discuss your data?

Incidentally, you cannot start thinking about your thesis soon enough. Whenever you conduct an experiment, ask yourself 'is this good enough for my thesis or a paper?'. Does the experiment have all the necessary controls, or have you relied on previous experiments to provide that control data? It is usually easier to repeat a properly controlled experiment, with samples and reagents that are working well, while you are familiar with and currently using the methodology. Similarly, you will be expected to show that you have tested basic assumptions about many of the methods that you have used, e.g. is the calibration curve linear? Over what range? Why not make sure that you have a thesis-worthy calibration curve prepared as early as possible; it will save you the considerable time you would waste in your final year if you had to resurrect the whole method just to prepare a 'pretty' calibration curve for your thesis.

Short-term time management

Within your research programme you will also need to manage your time on a weekly or even daily basis. In any week you will have to read the scientific literature, maintain your personal bibliography of that literature, design experiments, implement them and analyse/discuss the resulting data, attend seminars and perhaps teach undergraduates. How much time should you apportion to each of these activities? Here is another simple exercise: draw a pie chart that indicates how you see the distribution of your time on a weekly basis. Now, if your supervisor will humour you, ask him or her to draw, without seeing your chart, the same diagram for distribution of your time. Bear in mind that the pie chart cannot indicate the absolute number of hours in each working week, and thus conceals the fact that the area of your supervisor's chart would be about twice that of your own! Were there any major discrepancies in the two charts? Perhaps you apportioned different times to laboratory work, and too little to reading the literature? Discuss these discrepancies with your supervisor, and realize that it is a simple matter to let bench work expand to consume nearly all of the chart. Of course, at different stages throughout your research, the partitions will vary; here we are really discussing a long-term average. For example, you have to decide whether a week spent getting to understand and use a bibliographic database management system is an investment of your time that is recovered over the subsequent three years of your project.

Experimental design is critically important and you will save enormous amounts of time and effort by careful planning. A trivial point, but have you calculated the quantities of each material or reagent that you will require, and have you checked that there is sufficient of each? Will you have access to a piece of analytical equipment when you need it, especially if your experiment is time critical?

See Chapter 2, 'Developing your experimental skills'

Finally, take on board the truism that 'everything takes longer than you expect'. Always allow plenty of time for thinking, and for preparation. If you are preparing a poster or talk, remember that the construction of audio-visual materials is very time consuming.

The laboratory notebook

A laboratory notebook is a personal document that serves several important functions:

- it is the device for planning and recording experiments and data, on a day-to-day basis

- it serves as a long-term record of your work and permits extensive back reference to your work

- it is the source of information that you will need to write your thesis or papers.

  Provided that their scientific purpose is foremost, there is no reason why your notes cannot include 'smiley' faces in the margin for good experiments or self-chastisement for mistakes ('*What a ****** — I forgot to add enzyme!*'). It is a lasting record of your own research; a quick look at the scientists' notebooks in museums serves as a reminder that they can persist and be of interest far longer than we might ever have imagined!

During your undergraduate years you will have learned the basic principles of data recording and of preparing a written account of your practical work. However, postgraduate research training is very different. Published protocols that are used without modification are rare, and there is no direct assessment of written work in your note-books.

You should consider the type of notebook that you will keep before you begin your research, and make the decision in the context of the type of project that you will conduct. Most studies in the biological sciences generate substantial quantities of ancillary data items that are difficult to store; the correct recording and storage of these items is critical to your project, and you should adopt a system that will continue to be effective for the duration of your postgraduate research, and perhaps your whole research career.

Format   There are two common formats for laboratory notebooks. The most common is a series of hard-covered books, but a loose-leaf file format is also workable. Hard-back books have the advantage that the notebook is a single coherent document and individual pages are not readily lost. If you are pursuing several distinct aspects of a study in parallel then it might be preferable to use a different notebook for each aspect; it can be difficult to untangle the threads of parallel studies when they are intertwined in the same notebook.

The loose-leaf file is particularly useful when the project generates lots of ancillary documents such as photographic plates, spectrophotometer chart output, mass spectra or autoradiographs of blots or sequencing gels. These are always difficult to file and record correctly and a loose-leaf folder can be invaluable as a supplement to the notebook, if not as a complete replacement. Many documents can be punched directly for filing, but a useful alternative uses thin plastic bags that can be attached to loose-leaf sheets and used to hold nitrocellulose filters or photo-graphic plates. If a filter is susceptible to air oxidation (such as a western blot using peroxidase/3,3'-diaminobenzidine as a stain), it can be stored in a heat-sealed polythene bag. However, loose-leaf files are prone to one major problem — it is easy to remove individual sheets (for example, when preparing a manuscript) and lose them while they are extracted from the file. In the absence of page numbering, even re-insertion in the correct position can be a problem.

Index Irrespective of the format that you choose, it is essential that you compile, from the very beginning, some type of contents pages or index to your notebooks. You might anticipate producing between six and twelve notebooks during a Ph.D. project and finding your way around these after a period of two and a half years will be far from simple.

The simplest form of index is achieved by reserving the first few pages of the notebook for a contents section. Unfortunately, page-numbered notebooks are expensive and, probably, you will have to number the pages as you proceed. The contents can simply consist of a one-line description of the experiment, the page number on which it starts and the date. More complex cross-indexing might be complicated and hard to maintain and may not be worth the effort.

Indexing of ancillary data is critical. At the very least, each item should be annotated with the date and the location in the notebook of the corresponding experiment. Many items of computer-controlled equipment provide printouts of instrument settings and date/time annotation as well as output/analysis of the data — make good use of these facilities. Ensure that the clock on the computer is correctly set. Some computers still require that the date and time be set when the computer is switched on — do this! Most computers provide a battery-backed clock but make sure that it compensates, or that you do, for leap years and seasonal changes in time.

What to record? The throwaway answer to this question is very straightforward — 'everything'! In practice, you must consider the experimental design, protocols, data and analysis. There are two types of information in any notebook: objective data and subjective comments. The fact that you used 1 mM-EDTA is objective; the reason for using that concentration may be more subjective. If you have made a value judgment over the use of a particular technique, buffer solution, strain of bacteria or animal, etc., include your justification for this decision. Obviously, this does not need to be repeated at every instance, but such information will be invaluable in 18 month's time when you refer back to the experiment.

Further, the act of writing down your justifications will make you think more carefully about your decision and may prevent a wasted experiment.

Each experiment, or sub-section of an experiment, has a single independent variable; this means that all other parameters are defined. Again, recording the values of these parameters serves as a useful check on the experimental design as well as being invaluable for future reference. (You did **decide** to use those values of pH, ionic strength, chloride ion concentration or chromatography solvents, didn't you? They are **constants** rather than unwanted **variables** in your experiments, aren't they?)

Data must be recorded carefully. Resist the natural temptation to record the data 'in rough' and then transpose them to your notebook later — this is an extra opportunity for introduction of errors. When recording data from instruments, note the settings on the instrument panel. Three examples:

- a fluorimeter value of '10.4 units' has no meaning and cannot be rechecked unless you note slit widths, scale expansion factors, wavelengths and all other machine settings

- during an electrophoresis run note the current **and** voltage; this allows you to calculate the resistance of the gel and, hence, spot a buffer of incorrect conductivity

- in a chromatography run, note flow rate, column back pressure, detector settings, column type and, if there is more than one column of that type in the laboratory, the serial number of the column.

The act of recording this type of information is similar to the discipline of a 'pre-flight check' and can also prevent an unfortunate and sometimes expensive 'grounding' of an instrument!

It is difficult to be entirely objective about data. Sometimes a gel, autoradiogram or h.p.l.c. trace 'just doesn't look right', although you're not sure why. You must record this information in your notebook; in a few months time you may have done further experiments to explain the anomalous behaviour. Don't be afraid to make subjective comments; the response of the human intellect to data, especially unexpected data, underpins the development of research. A set of comments added to your notebook *post mortem* are a great help in clarifying your analysis and in helping you design the next experiment.

An 'ideas' notebook?

This suggestion is very much a matter of personal taste. Consider keeping a rough notebook that you use for jottings and notes, as you think about and plan your project. Use it during discussions with your supervisor to record your collective thoughts. This might be the one that gets into the museum! At the least, is it good to have a brief summary of your meetings with your supervisor; good ideas, discussed in passing, can be lost in a high-paced meeting.

Storage and security

Try this (slightly hazardous) experiment: go to a final-year graduate student (they are recognizable as confident and authoritative individuals working at a great pace), and ask them gently what would happen if their accumulated laboratory notebooks were destroyed at that moment. Ask them quietly where they keep their notebooks at night, and whether this location is protected from the ravages of fire and flood. Note the correlation between the change in facial expression and the answer:

- relaxed and serene 'My copy is written in water-insoluble ink and stored in a heat-resistant, waterproof safe; my supervisor's copy is kept at home.'

- perplexed and defensive 'There they are, that pile on the window sill, except the one I lost when my briefcase was stolen. Yes, they are a bit crinkled, that was the flood from Lab XXX last year, but I think I'll be able to separate the pages...it's a pity the blue fibre-tip pen ran so much though.'

It is ironic that many laboratories seem to give more consideration to the storage of reprints, which are copies of existing literature than to notebooks, which are irreplacable originals.

Two exaggerated extremes to emphasize that you must consider the security of your notebooks. Duplicated notebooks, whether as carbon copies or photocopies, are costly to maintain, and the majority of supervisors will not select this option. This decision puts additional emphasis on the storage of your notebooks — the only record of your efforts. A steel filing cabinet gives good protection from indirect heat or water; if you do not have access to such a cabinet, you should make a request for this facility.

Storing data in machine-readable format

The nature of biological data is broad-ranging, and there are some types of data, such as nucleic acid sequences or structure coordinate files that only really have meaning or value within the context of computer-based methods of analysis and data representation. Additionally, it is becoming increasingly difficult to purchase any piece of equipment of some sophistication without an obligatory computer that provides both control of the instrument and collection, analysis and storage of the data. A student might therefore be inclined to ask whether the best method to store all experimental data is in computer-readable form, and whether hard-copy (paper, human-readable) representations of the data are necessary. The answer to the latter is undoubtedly 'yes'; computers do not make margin notes, or draw a box around an interesting feature in a sequence. However, machine-readable storage of raw data or of the subsequent analyses is an option that should be considered.

What data should be stored?

If you elect to store and archive some of your data in computer-readable form, you will have to make decisions about the format of the stored data. In general, the maxim should be 'maximum information, minimum file size', but this may be qualified by other considerations. To illustrate: consider a scenario in which you prepare two typical northern or western blots (a test and a control sample), scan the resulting autoradiographs in a densitometer, and save the scans on disk. The scans will occupy somewhere in the region of 10 Mbytes on the disk. From the scans you subsequently derive the intensities of 12 different bands or spots, and store the quantitation data in a separate file which will be approximately 250 bytes long. The size ratio of the original data to the analysed data is 40 000 to 1. Since the original scans will not fit on a floppy disk, they will have to be stored on a hard disk and, by current standards, even a large 250 Mbyte hard disk will only be able to store about 10–20 images. Does the huge size of the digitized images warrant the space they occupy on disk? If the sample is especially valuable/irreplaceable, or the results of great significance, then there is a stronger case to be made for retaining the file. It may be possible to make a copy on magnetic tape, which can accommodate large files.

Nucleic acid sequence files are almost impossible to handle and analyse without the assistance of computers. Most sequence analysis software allows the user to save the results of the analysis on disk, and the space required for these analyses will usually be much larger than that required for the sequence itself. If the analysis can be repeated readily, there seems less value in storing the results than if the analysis was conducted in such a way as to make repetition difficult. As in most of these examples, data storage and file maintenance must be weighed against value and file size.

It is likely that you will generate graphs of data sets by graph-drawing software, and that you may use this software to generate an image of the data for your notebook. Most of these packages allow you to save the graphic image of the data set to disk in addition to the data themselves, but you should be aware that the file size of the graphical images is often much larger than those for the original data sets, and you will waste disk space on images that can easily be recreated. Save the data, but worry less about the pictures; print out the graphic for your notebook and then discard the graphic file.

Formats for data storage

Machine-readable data can be stored in a range of different formats, and you should consider the formats that will be of most value to you now, but also in the future. The three years of a research project is a long time in the development of small computer systems. In the final stages of your research programme you may wish to analyse data that you generated in the first few months of your research, using software that might only have become available recently. What guarantee have you that the data you saved to disk so long ago is accessible to the latest software? Unfortunately, the answer is very little or no guarantee. Indeed, if the data are few, it might be preferable to re-key the data into the new software, which is another reason for making sure you retain hard copy of the data set.

ASCII — American Standard Code for Information Interchange. A code used almost universally for data exchange. Each character of a text file is represented by a code, such as ASCII 32 (space) and ASCII 65 (A).

There are some steps that you can take to ensure longevity of your data. Many software packages store data in several formats, ranging from internal, efficient formats (the most specific) to simple text files with little additional information (the least specific). The latter is likely to be readable by almost any program, the former will only be readable by software that recognizes that specific format. If your program uses a very specific format, you might be advised to save the data also as a simple text file. This is often provided as an option in software, and is sometimes referred to as a 'text file' of 'ASCII format' data. In this format, each digit of a number, or base of a sequence is stored as a character, and the only non-printing characters that are usually found are the 'tab' character (separating columns of data) and the 'carriage return' character (signifying the end of a line). Almost any software package should be able to accept this data.

Security All media for computer-based storage of data will fail at some stage, and it seems difficult to predict when this will occur. The 'lifetime' guarantees offered by some suppliers of removable disks (*'if the disk fails, we will replace it free of charge'*) may offer some indication of the inherent reliability of the medium, but a blank, brand new disk is no substitute for a disk full of data from unrepeatable experiments. Poorly written software, or the simple act of removing a disk while information is being written to it can corrupt the entire disk. Viruses, small programs that deliberately and malevolently damage other files can have devastating consequences. You might find it easy to detect a virus that erases a data file or disk, and you may be confident that your back-up copy of the data is fine, but how would you spot a virus that sporadically changed occurrences of '9' to a '0' in your data files?

See Chapter 6, 'Computers and computing'

The more data you store in machine-readable form, the more vigilant you should be about data security. Most computers offer a fixed ('hard') disk that can store a great deal of data, but you should never rely on a hard disk as your sole source of that data. Hard disks are usually used by several people, one of whom might accidentally manage to delete every file. Public hard disks tend to become littered with many files, most of them untraceable and with unmemorable filenames such as 'RUN4.DAT' or 'CONTIG.SEQ' and they may be periodically expunged of such files. A hard disk, unless it is available virtually for your exclusive use, should not be relied upon for long-term storage.

Removable ('floppy') disks have a much lower capacity than hard disks, but because they are under your complete control they can be used exclusively for your files, and taken away and stored in a secure place. Keep at least two copies (both on floppies, or one hard disk, one floppy disk copy) of any files, in different locations.

Organize data in a manner that facilitates easy backup. Store your data in subdirectories that will fit on a single copy disk; you can copy all the files within that subdirectory on to a floppy disk with a single command. Learn the basic commands of the disk operating system/file management system on your computers.

Floppy disks are cheap by comparison with the cost of the time you have invested in acquiring the data. Worry less about cramming files onto the disk than about organizing your data in a logical fashion. A final trivial point that is often overlooked — label the disks clearly with their contents, your name and address or 'phone number.

All of this discussion has been focused on the physical security of your data, and has not addressed the other aspect of security; namely that of keeping data from the eyes of others. If data security in the 'espionage' sense is important, then you should be appraised of this in advance and be given strict procedures to follow. Thankfully, in the great majority of laboratories, this is not a significant issue.

Keeping up with the literature

It would be quite possible for you to occupy the whole of your Ph.D. project in the library, ploughing through the mass of material that is published every year in the biological sciences. Thousands of scientific papers, hundreds of reviews, dozens of books will all have some connection with your work. Some of them are of direct relevance (it is distressing, but rarely terminally tragic, to see your own research topic published by others!). Yet, your aim is to contribute to that body of knowledge by publishing the results of your own endeavours. How then can you reconcile your need to spend the majority of your time in the laboratory, planning your own research and yet keep up to date with related work as it is published? The golden rule is be selective. There are two basic strategies for keeping up with the literature — by manual methods, or by computer-directed searches.

The only way to cope with the scientific literature is to be highly selective in what you scrutinize and read!

Manual methods

Manual methods require that you devise a search strategy and then use that strategy to discover publications of interest. The simplest system is where you elect to monitor a limited set of journals that are the strongest in your discipline. You will fail if you attempt to keep up to date with every journal. A typical list might comprise between 20 and 30 journal titles. This list will evolve and change as your programme of research develops, but in the first instance your supervisor and others in your laboratory will be able to advise you on your list.

Having established this list, how do you go about keeping track of the material? Two methods are commonly used. You could read through the journals as they are received by the library, in the current periodical area; this is obviously restricted to those journals to which your institution subscribes. However, it is all too easy to slip behind, and eventually unseen journals will disappear for binding. Advantages are that you read the whole contents page of the journal and thus 'browse' in the subject, and also that you check, by quick scrutiny of the article itself, whether it is worth further reading.

If you have to use this method, do not let back issues mount up, as the task of reading them quickly attains a magnitude such that you will not bother.

More likely, you will wait for the journals to appear in a document produced by an abstracting service, the best known of which is *Current Contents*. This comprises a photographically-reduced version of the contents page of many journals, it collates authors names and addresses and cross-references the citations. This is the method favoured by many but has two drawbacks. First, there is a significant delay (measurable in weeks to months) before a particular issue of a journal appears in *Current Contents*. Secondly, *Current Contents* generates a list of papers, but you must still go the library to view these in full before you can assess their true value. However, *Current Contents* does allow you to browse journals that are not in your library.

When you write down a citation, for example, from *Current Contents*, make sure that you record, accurately, enough information to allow you to find the paper afterwards. At the least, record the first author's surname and initials, the first few words of the title, the year of publication, journal name, volume and issue, and the first and last page numbers of the article. Even if some of this information is redundant when you come to look for the article, it will all be needed if you have to complete a reservation slip or request the article by inter-library loan.

Having located a paper of interest; what should your next move be? Ideally, you should write to the author, who will send you a reprint. In practice, this is very inefficient. The only real solution is to use a photocopier. Everybody has at some time photocopied a paper that subsequently they have never read. Avoid the temptation to go on a photocopying spree. For many papers, a single page containing the title, authors, full citation and abstract will provide enough information, when combined with your own notes on the paper. The copy of the abstract can be cut out and stuck on to a record card if that is the method that you choose to maintain your bibliography.

Incidentally, you may wonder about your position with regard to copyright legislation when you copy an article. You are usually allowed to make a copy of a paper for personal study, but you may be required to provide some written record of the article that you copy, such as the ISSN number. Your institutional librarian will give you advice on this matter.

Computer-based methods for literature surveys

Keeping up with the literature is a task ideally suited to computerization. Given access to a database of the scientific literature, recently published or published sometime in the past, it should be possible to extract from that database citations that are of direct relevance to your work. Moreover, a search according to an author's name can identify related work, or provide a good indication of the work and productivity of a potential future employer!

*As a forward-looking life scientist, you would need a very good reason **not** to use computer-based methods!*

There are several types of database that you could use. *Current Contents* is also produced in a computer-readable form. As with the paper version, it is published weekly on several floppy disks. Each issue must be assembled on to a hard disk, and then searched using the *Current Contents* searching software. Searches can be performed according to keywords, author's names, etc. or individual journals can be browsed. A more detailed version of *Current Contents* (and one which comes on over twice as many floppy disks each week!) includes the abstracts of most articles. The abstract can be browsed to assess the importance of the article, and can also be searched for keywords. With either database, articles that you identify as being of value can be moved to a personal interest checklist and from there can be printed out, to take to the library.

Every week, you might typically expect to discover five to ten articles of immediate interest and perhaps the same number or more of peripheral interest or covering methodology. You will read the most important articles, but you are unlikely to copy them all. Yet, you will want to retain some pointer to these articles for future reference (e.g. when writing your thesis). By all means keep the checklists as they are printed out, but even these become tiresome to look through. Preferably, transfer the references to the articles, as they accumulate, into a personal bibliographic database.

See Chapter 6, 'Computers and computing'

Often, you will want to look back through the literature for work that has previously been published in a particular subject area. It is impractical to browse old issues of *Current Contents* as these will most probably have been discarded long ago. Rather, you need to search a database of all published literature in the biological/biomedical sciences. Until recently, the size of these databases meant that they could only be stored on large mainframe computers, and they were accessed remotely by computer terminal. Charged by connection time and the number of citations extracted, these databases did not encourage browsing or the acquisition of skill in searching. Now, the development of CD-ROM technology means that such databases are within immediate reach of all libraries, most departments and even a few individual laboratories. Also, new resources, accessed through computer networks, are appearing daily.

Your library will definitely have CD-ROM technology, and should maintain subscriptions to the most popular biological databases, such as Medline (biomedical journals) or the Life Sciences collection. At the beginning of your project, book a training session with the librarian or information specialist, and become competent in the use of the system. Almost certainly, you will only have access to the CD-ROM system for short periods, and you should use this time to explore the database and capture as many citations as you can in the allocated time. Do not waste time printing the output — find out how to transfer your results on to a floppy disk, and take the references away with you so that you can print and browse them later at your leisure.

Of course, the output of a database session is of least value to you on paper. If at all possible, discover how to output the results of your search in a form that allows easy import into your personal bibliographic database manager. You will need to know what formats are expected by the database manager, and what formats can be produced by the retrieval software; pick formats that are compatible.

Typical search term: ((zinc and (metalloendopeptidase or metalloproteinase)) not thermolysin)

Searching databases by computer can be complex, and it is easy to get 'lost' in the database. Learn the difference between the logic terms 'and', 'not' and 'or'. Although these are words you use every day, their meaning in database interrogation is much more precise.

Some final words of advice. The scientific literature is moving faster than you at the best of times. Don't let this aspect of your research lapse. It happens easily and you must be vigilant in your attempts to keep up, even if seems as though you are making little headway. Secondly, please **read** the papers that you discover. It is too easy to believe that the act of photocopying a paper is enough. Take the copy home, read and digest it slowly, think about the conclusions, and how the work impinges on your own. Is the methodology such that you could design a different experiment in your own research? Incidentally, don't be afraid to bring an interesting paper to the attention of your supervisor. There is a chance that they are behind you in their reading, will be grateful to see the article, and will be impressed by your diligence. Most importantly, you will be doing your part in moving the research programme onwards.

Maintaining a bibliography

When you start your research programme you will amass a pile of reprints or photocopies of papers. Initially, they will be so few that you will be able to keep track of them simply by sifting through the pile. Then, as their numbers start to swell it will become increasingly difficult to keep them under control; they will start to migrate to far corners of your department, your home, your supervisor's office, behind the centrifuge.... Moreover, you will come across some material that you might not elect to copy, but would nonetheless like to record for future reference. Inevitably, you have to face the daunting fact that you will need to set up a personal bibliography. No matter how daunting this might be, remember that the larger your reprint collection the more difficult this will be, and the sooner you start the better.

A bibliography for a postgraduate research programme has a number of functions. It acts as a compact and localized index to the literature that you have identified as relevant to your project. It can allow cross-referencing by subject or author and allow you to trace a citation very quickly (however, this will be followed by much ranting, raving and disturbance of the contents of your desk, because finding the photocopy will not be so straightforward!). Also, if your personal bibliography is constructed correctly it will make the task of organizing the references for your thesis a lot simpler; either re-ordering a pile of reference cards and handing them to your typist, or pressing a couple of keys to generate a corrected formatted bibliography, sorted in whatever order you need.

Manual systems

There have been all sorts of systems to allow a bibliography to be manipulated on index cards. The underlying principle of a card-file system is simple. The card index is compact, there is no need to photocopy every article, it doesn't take long to sift through a card index of Ph.D. project proportions and the cards are readily hand-written. However, the cards can be lost, shuffled or soaked, and are readily borrowed by a colleague who wants to look up a citation in the library. Writing the cards takes time, although writing out a citation can be useful as mental reinforcement.

Again, you should really seek to use a computer-based bibliography from the outset.

Undoubtedly, the biggest drawbacks of a card-file system are the time needed to maintain the system and the organization of the cards within the system. Should the cards be filed into sections that are subject-oriented? Or should they be filed alphabetically by author, chronologically by year, or chronologically in order of discovery? If you use any of the non-subject filing schemes you will have to assign each card a unique reference number that is subsequently cross-indexed to other subject or author indices. If you use subject-based filing, what happens when you come across a card that should be filed under two subject headings: write it out twice or insert a card saying 'see also Bloggs *et al.*, filed under ...'? Undoubtedly, any of the schemes can be made to work, but their maintenance requires diligence and care. Since such cross-referencing, indexing and rapid searching are natural functions of the computer, should you not learn how to use a database management system or a bibliographic package?

Computerized personal databases

Storage, sorting and manipulation of records of a database is ideally relegated to a computer. First, some jargon. A simple database consists of **records**, each of which is equivalent to a single citation. Each record consists of **fields**; these might for example include authors, title, journal, volume, first page, last page, year and abstract. **Indexed** fields are easy to search, non-indexed fields may be slow to scan through.

Consider the space needed to store a single citation. Many database management systems (or DBMS; the program that creates and accesses the database) require that you specify, during the creation of the database, the amount of space that you will assign to each field. The space can for the moment be calculated in alphabetic characters but computer aficionados can assume that one character will need about one **byte** of storage). In 1992 one paper in *Nature* had 147 authors. Allowing at least 10 characters for surname and initials, plus a few extra for punctuation, you would therefore need about 1500 characters of storage for the biggest author field conceivable. Yet, in most citations, there will be between one and four authors and most of the 1500 characters would be unused. This indicates the value of a DBMS that allows **dynamic** storage of field contents so that fields can contain, within reason, any amount of material without wasting space in other, more compact records. Another advantage of dynamic space allocation is that you do not need to specify the size of each field in advance.

However, even with dynamic allocation of space to field contents, a typical literature citation cannot normally be contained in less than about 500 characters if the title is included. If an abstract is included as well, the storage requirement can increase to about 2500 characters per record. The records must be stored on floppy or hard disk. The database can now be as large as the disk (comprising thousands of records). Disk-based DBMS work well and give rapid access to a large database.

There are two options in adopting a DBMS to maintain a bibliography. You can design your literature database from scratch, using a general purpose DBMS that allows you to specify the fields that you want, and decide which of them are indexed for rapid searching and sorting. Preferably, you should use a customized database management system that has been designed specifically for management of bibliographies. This type of system is usually well-tailored to the needs of an individual scientist and offers facilities such as the ability to add index terms, keywords, and an abstract, as well as allowing the printing of selected records in a format that is compatible with the house style of a particular journal (or according to your thesis requirements).

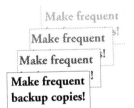

If you use a DBMS, one 'rule' to bear in mind is that the security of a database is inversely related to the number of records that it contains. The cat will never destroy a disk with only 10 records. It seems to know, and will wait for the time when the disk holds 250 records and you haven't a duplicate anywhere...

If you work in a laboratory where there are several scientists with related interests, it is preferable to develop a laboratory-wide bibliographic database. This will then function as a resource for research, for thesis and paper writing, and is superior to a situation where individuals each develop their personal bibliographies that overlap by about 70%

with each other. If the subject matter is sufficiently focused, the database can be updated by importing records from a search of the major literature databases or from a weekly *Current Contents* search; it would normally be possible to include abstracts in such a database.

Thus, if you are lucky, you may have access to a laboratory bibliographic database that already contains thousands of records. How then can you select the two to three hundred that you will eventually want to place in the bibliography of your thesis? Clearly, you must be able to select those records that are particular to you, which means that somewhere in each record you will have to add a 'string' of characters that indicate that this reference is destined to appear in your thesis. The string might be as simple as your initials, but beware of 'false positives' if your initials are 'PCR' or 'DNA'! It is preferable to use some type of unique identifier, such as "**RJB**" Also, colleagues in your laboratory will not be pleased to discover your unique string in the author lists when they prepare the bibliography for a paper. Most bibliographic databases will include fields that are not normally printed in the final bibliography, such as keyword or note fields. The keyword field usually contains a list of terms and your identifier can easily be added to that list. Then, extracting your thesis/paper references is a simple matter of searching for your string, sorting the records that match that search, and formatting the output of those records in the style that you want. The database is not modified, you do not create a subsidiary database and you do not affect the ability of others to use the database.

Attending lectures and presentations

During your graduate studies, you will be given many opportunities to attend seminars and lectures, ranging from single departmental lectures and one-day meetings to week-long meetings that are either specialized or large international meetings that cover many different subjects. How you approach these lectures is very much a personal matter, but a little forward planning can make a big difference to the benefits that you may glean from them.

Which lectures should you attend? The biological sciences are increasingly multidisciplinary, and it is difficult to justify missing a lecture on the basis of relevance to your field. In reality, there will be very few lectures that match your area of research, which is just as well, because you will not want to see your thesis work published elsewhere. Departmental seminars are likely to be broad-ranging and are a good opportunity to extend your range of interests, and to develop your understanding of other research techniques. Attendance at departmental seminars may also be a requirement of your graduate study programme, and is in any case a courtesy to be offered to the visiting speaker.

Scientific meetings are rather different. A one-day programme focused on a single subject can provide an excellent update in an area, and most scientists will be able to sit through all the lectures in the day's programme. However, when meetings extend over a few days you will find that your powers of concentration are tested severely and unless you are superhuman, or make careful plans, they are likely to wane to nothing. Some meetings have morning, afternoon and evening sessions with perhaps three or four lectures per session. After five days of such intense activity, you will probably have difficulty remembering what day it is and what country you are in! Be selective. Use the meeting abstracts to identify those lectures that you will attend. Aim to attend all plenary lectures — these will usually be given as broad overviews by acknowledged experts — and all lectures that are particularly relevant to your own research. From the rest, pick the lectures that you would like to attend for other reasons; perhaps you know the speaker, or want to find out more about a technique. Never feel embarrassed at leaving the lecture room in the interval between two lectures, although it is discourteous to leave in the middle of a presentation.

Many scientists seem unable to shrug off the habit of their school/undergraduate days, and float to the back of the lecture room. Leave them at the back, and occupy the front rows. You will see the slides better, will have your attention sustained by the eye contact with the lecturer and, in a banked lecture theatre, it is usually cooler and more comfortable. It is easier to ask a question from the front of a lecture room whereas projecting your voice from the back of a large theatre is difficult. Should you take notes? This is very much a matter of personal preference, but as an undergraduate you will have discovered that it is difficult to listen, write and understand simultaneously. It may be better to concentrate on understanding, making a few notes of key points which have particular relevance to your work, or of ideas and experiments that might suggest themselves to you as you draw parallels with your own work. You might record a point that you did not follow, to remind you to seek clarification from the speaker later.

*See Chapter 5,
'Talking about your
work'*

There are other games that you can play in lectures. Assess a lecture for the quality of the presentation, as well as the content. Were the slides/ overheads legible? Rate the speaker, and then ask yourself why you thought the talk was good or bad. Promise yourself that you will emulate the good features and avoid the bad features of these presentations. At question time, enjoy the interactions and game playing that goes on between the chairman, speaker and members of the audience. Learn how to ask, and how to answer questions; your time will come. If you want to ask a question yourself, plan the sentence, or even write it down.

Last, but not least, meetings provide an opportunity for informal interaction with other scientists. At poster sessions, in pre-prandial social events and over meals, you will have the chance to speak to many of the attendees. Do not be put off by the 'greatness' of the big names in your field; they are usually notable scientists because they enjoy science, and you will find them more than willing to talk to you, the next generation of scientists. Introduce yourself, make some introductory remark about your work and ask them your question. You can be virtually certain of courteous attention, gentle questioning and good advice.

Predoctoral meetings

The more you put into attendance at a meeting, the more you are likely to benefit from it. Several learned societies provide opportunities for graduate students to attend meetings at which they will be virtually the sole attendees. This is an ideal opportunity to practice your communication skills, whether as a member of the audience or as a presenter of a talk or a poster. You should want to make the most of these opportunities; they allow you to talk about your work and to discover the work of others in a friendly, yet scientifically rigorous, environment.

Who pays?

Of course, attending meetings costs money and, for graduate students, there are relatively few sources of funding that can be explored. Specific details cannot be given, as individual circumstances vary so much, but the following guidelines might help. Most importantly, your chances of being funded to attend a meeting are considerably strengthened if you intend to present some of your own work, whether in the form of a poster or oral presentation. It is helpful to include the abstract of your presentation when you apply for travel grants. Meetings organizers are usually aware of the funding difficulties experienced by young researchers and will often provide bursaries that subsidize the meeting costs.

These are often limited in number and extent of subsidy; it is in your interest to make enquiries as soon as is possible.

Even with a bursary, you will still have to find the remaining funds from other sources. In addition to the body that funds your studentship, there are many charities, learned and professional bodies that are able to make travel grants available, although not all of them will provide funds for predoctoral scientists. If you are involved in a project that has an element of industrial sponsorship, the collaborating company might be able to help you attend the meeting as they often have an earmarked budget specifically for meeting attendance that might be accessible to you. Again, make enquiries as early as you can. If you make application to more than one funding agency, it is a courtesy to make this known to each. The converse also applies; if you apply to a single body then you should make this clear in your covering letter. Often, your application will require a letter of support from your supervisor. You will be asked to estimate your travel costs. Be realistic, and include all travel costs, subsistence costs and registration fees. If you can take advantage of any cut-price travel deals then do so; the lower your costs the more likely you are to be given support.

If you are successful, write to your contact at the funding body after the meeting, thanking them for support, and indicating the value of the meeting to you. If you are awarded more money than you need, return the unused sum, in proportion to the amounts initially awarded by the different funding agencies. These courtesies can only help to pave the way for your successors.

Professional bodies and scientific organizations

As you embark upon your scientific career, you should join one of the learned or professional bodies that is particularly relevant to your research area. Your supervisor will advise you. For your membership fee, which is usually very low for graduate students, you will receive regular notices and programmes of meetings, newsletters and other information. You will be able to attend the meetings organized by the society and may be able to apply for travel bursaries to attend them. You will start to discover that there is an organizational and slightly political side to science, and will learn to recognize some of the key players. This is all part of the process of becoming involved in the full scope of your chosen science.

Writing about your work

Introduction

Regardless of its importance or value, your work has little meaning unless you can communicate it to others. In nearly all instances the medium will be the written or spoken word — you are unlikely to sell the film rights or make a video of your thesis! As part of your development as a professional scientist you must acquire skills as a communicator, and become as effective at writing and speaking about your work as you are at doing it. Most likely, you will invest little time in this part of your training, will find the skills more difficult to acquire, will not know when you have attained an adequate level of competence and will miss the feedback and reassurance that you get from a beautifully executed experiment. Some scientists are superb writers or speakers, and a few are downright terrible — at this stage your goal should be to avoid emulating the latter, and start to work towards the former. A great deal has been written about communication in science, and you should seek out and read these books in your library; the reading lists are a useful place to start. This chapter and Chapter 5 can only give some general advice.

See Chapter 5, 'Talking about your work'

During your undergraduate years, you probably had little or no guidance about the skills of effective writing or speaking in public. You will have to learn these skills, just as you would any laboratory technique, by reading, practice, and through advice and criticism. Your ultimate goal is to convey clearly, and without ambiguity, the scientific message. You should seek to develop a style that conveys your personality, as well as acting as an effective vehicle for delivery of your observations and interpretations.

This Chapter, devoted to writing skills, focuses on the work that graduate students are most likely to have to produce. Many of the suggestions can be carried over to other documents that you might have to produce in the future. During your studies you will be writing brief abstracts, and at least once, you will write a major piece of text — your thesis. At intervals throughout your training you may be expected to write progress reports, summaries of your work, and maybe applications for travel funds. It is unlikely that you will be expected to write papers unaided, and very unlikely that you will write research grant applications.

Unless you are clear about what you want to say or write, your communication is going to suffer, both in content and structure. You must know the background and your experimental system very well. The need for planning cannot be overemphasized. The content, the logical flow of ideas, the duration/length and the use of appropriate graphics to convey information must all be considered. A shortage of time will pressure you into errors or taking short cuts: plan these exercises well in advance. Graphics are used extensively to convey scientific data and models and you must be proficient in the design, preparation and use of these visual aids.

Effective scientific writing

How do you learn the skill of effective scientific writing? You will read thousands of words every day. Every so often, take time out to consider the writing rather than the content. Did the writer communicate with clarity and elegance; if so, what was it about a particular piece that appealed to you? Take one of those sentences that needed several readings before its meaning became clear — could you have made a better job of it?

Read one of the many books on scientific writing. Some are rather dull and dry, but others are excellent: pithy, helpful and witty. Some suggestions are given at the end of this Chapter.

Invite criticism

Offer your work to your supervisor and invite comments on the writing style and grammar as well as the scientific content. Be prepared for the depressing sight of a page covered in editorial 'blue pencil'. Read over these comments, and establish whether the highlighting has been reserved for grammatical errors or whether sentences have been restyled. Your supervisor has benefited from a longer experience of scientific writing and will suggest alternative ways to express your ideas. These may appeal to you, but analyse the reasons why you feel that your supervisor's style is better than yours. Try to carry the principles into your own writing, but at the same time try to develop a style of your own.

An additional responsibilty

There are many whose native speaking language is not English, the primary language of the international scientific community. English is not easily learnt and native English speakers should be rather humbled by the linguistic skills of those who succeed in communicating in what is for them a difficult foreign language. As an English speaker, you should appreciate the need to avoid colloquialisms and to aim for clarity and simplicity in your written work.

Produce an outline

How do you start? Unless you are very unusual, you will not begin with the first word and write the whole piece to the end. That presupposes that you have a clear view of the structure and content of the piece at the outset, which is rarely the case. A plan or outline will clarify the logical structure that is found in all good scientific writing. This exercise can take many forms, but always start with a clear mind, a figuratively empty desk and a blank sheet of paper. You will not make a good job of your plan if you are constantly leaving your desk to take samples during an experiment. Nor is this the time to be looking up precise details in your notebooks; at this stage you should be interested in the overall ideas and the inter-relationships between them:

— what are the concepts that you want to convey?

— how much space (words) do you have to express those ideas?

— how much background information must you provide?

— what data will you need to support your hypothesis?

— how will you present the data?

Your plan can consist of major ideas, written down as boxes or balloons as they come to you, and connected by lines to other related ideas. Or, you may prefer to use an indented heading/sub-heading structure as an outline.

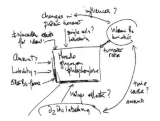

When you have the overall scientific content clear in your mind, break the work up into sections. Usually, this is decided for you by external requirements, and you will plan much of your writing in the form of abstract, introduction, materials and methods, results and discussion. At this stage, it is sometimes helpful to expand your original plan, and now use a sheet of paper for each section. Again, write down everything that you want to include, and keep these sheets for new ideas or material that occurs to you as you are working on the text. Perhaps you forgot in your original plan that you would have to include a description of a method that you modified?

Organize your material

Except for all but the shortest pieces, file the materials, including outlines and any floppy disks containing text, in one place, such as a folder or wallet (you will of course have backup floppy disks elsewhere?). Mark it clearly with the subject, your name and telephone number. If you want to use data from your laboratory notebooks, consider putting photocopies of the material in the folder; this can prevent a great deal of distraction when you need to refer to that material. Keep previous draft copies with margin notes and corrections until you are sure that you no longer need them. Always keep one paper copy — in the even of loss of a disk file, it is easier to retype a paper version than to recreate the manuscript from scratch.

First draft

At this stage, you are in a position to produce the first draft. Your main objective should be to produce a document that captures all of your ideas in a logical structure. Worry less at this stage about details such as references, Figures and Tables. Try to use paragraphs to organize ideas. Use the headings and subheadings that you will employ in the final version.

When you have written a first draft, give a copy to someone to read, and put it out of sight, and out of mind, for a week or so. Then, before receiving any other comments, read it yourself. You will be surprised at what you have written, and will want to make many improvements. Invite the comments of the person who read the text, and make sure that they understand that you will not be offended if they are strongly (but constructively) critical.

Style and grammar

In the early phases of your writing, almost every sentence can be improved by rewording. Reorder ideas or subjects, use alternative or simpler words or break a sentence into two. While you read over your first draft, begin to consider style and grammar. Look out for phrases that you use, but should avoid. A common one is found to be 'found to be'; this can usually be struck out of the text without altering the meaning, and tightens up the prose substantially — *try re-reading this sentence omitting the first 'found to be'*. Another is the construction 'Bloggs *et al.*, (1993) have shown(/found/discovered/demonstrated/ established/indicated/proved) that X = Y'; replace it with 'X=Y (Bloggs *et al.*, 1993)'. There are many more of these *bêtes noires*; try to identify your 'favourites', and expunge them from your writing.

Learn how to use and apply the hierarchy of separators: paragraphs, full stops, colons, semicolons, hyphens and commas (see Booth, 1985 for a brief but clear discourse on this hierarchy). You must assume the responsibility for the production of text that conveys its message elegantly and efficiently.

The teaching of grammar as a formal subject seems to be out of favour, but that does not absolve you of the need to produce grammatically correct text. Read Booth (1985) and O'Connor (1991) for concise and readable discussions of elements of grammar and style as applied to scientific writing.

Second draft

The second draft will often refine ideas and structure, and develop style and grammar. Introduce the first versions of the Figures and Tables, and relate the text to the data. References can be added, and the details missing from the first draft can now be included. At this stage you have a document that is starting to resemble the final product.

Final version

If you have detailed instructions for production of the document (such as 'Instructions for Authors' from a journal, or departmental guidelines), read them carefully, and comply with them. There may be cosmetic changes and subtle refinements at this stage, but eventually you will be in possession of the finished document. If you still have time, put it aside and read it again, from beginning to end, in a few days time. It is difficult to read a document and assess content, logic, style, grammar and format all at the same time, and you may prefer to go over the text several times. Be prepared to make changes even at this stage despite the natural tendency to resist changing an attractively produced, near-final version. But, at the same time be aware that repeated alteration of a manuscript reaches the stage where little is being achieved. It is reasonable to expect that your writing skills will improve from one document to the next, and you should not be aiming for perfection in your first manuscript.

Part of taking responsibility for your studies is leaving adequate time for production of written work to deadlines.

Word processing

It would be surprising if you do not have access to word-processing facilities. If you have a choice, learn how to use a 'mainstream' package that is likely to be updated and improved during the course of your research project — this may mean a little background reading and research before you start. If you do use a word processor, make sure that you are familiar with its use before you have to produce material to a deadline. Make absolutely sure that you know how to save and retrieve the latest version from disk. All word processing software comprises two parts, often combined seamlessly: one part for text editing, the other for page layout. The text editor should make it easy for you to produce successive drafts of your work. Search and replace facilities, spelling checkers (to correct typographical rather than spelling errors, of course), mouse and/or cursor key movement through the text

See Chapter 6, 'Computers and computing'

should all be obligatory. Desirable features include the ability to define text styles and glossary items. Some packages even offer to report on and correct your grammar. The page layout facilities should include the ability to incorporate graphics from other programs, and easy control of margins, footnotes and running headings. Make sure that the package can send output to all the printers you anticipate using, from simple dot-matrix devices for drafts to a high-quality printer for the finished versions. If on-line help is included in the package, learn how to use it.

Abstracts and summaries

First-year reports, second-year poster presentations, theses, papers and presentations at external scientific meetings will all require that you produce a précis of the contents of your presentation. The production of a short abstract or summary is often dismissed as a trivial task. Remember though that the abstract is often the only part that is read by the non-specialist. It will be used to help other scientists decide which papers to read, which lectures to attend and which posters to visit. This short piece of prose, usually about 200–500 words long, must serve to inform the non-specialist and attract the attention of the specialist. It should carry a brief description of the important conclusions and their bearing on the field.

As with all writing, this short text will go through several iterations, so in the early stages do not worry unduly about length. Concentrate on the content, making sure that the points you wish to emphasize are included. Then take a critical look at those points. Do they really reflect the rest of the text, or the presentation, or have you concentrated on peripheral or obscure points? Have you made claims or exaggerations that are too bold for the data to sustain? Have you wasted space by letting the reader know that 'data will be discussed'?

Having decided what to say, express these points in a short piece of careful and lucid text that is interesting to read. Hand it to a colleague who is not particularly familiar with the work and request their opinion on its content and style. Ask them what they would expect to find in your report or other presentation. Be prepared to rewrite.

Many abstracts and summaries are photoreproduced directly from the typescript as supplied by the author to eliminate the need for a proofreading step and to expedite publication. This is often referred to as camera-ready copy. Often, a form will be provided that contains a blank space in which the abstract may be typed. In general, it should not be necessary to type directly on this form. Measure the typing area carefully and type or print to the same dimensions, check the abstract and finally cut it out and stick it into the space provided. This reduces the possibility of a ruined abstract form. Often, you will not need to use the form at all, provided that you use the same dimensions as in the typing area. If in doubt, ask.

A high-quality laser/inkjet printer will give the best printed output, but use a font that is generated at printer resolution, not screen resolution. If your abstract is prepared by typewriter, use a good carbon ribbon to give a strong contrast on the paper. When the abstracts are reproduced photographically, a faint image will lose contrast in this process. Never use a low-resolution dot-matrix printer as the ribbon never gives a good enough contrast, even when new. Abstracts reproduced from such printers fade into obscurity and do little to encourage a colleague to pursue your work. NLQ (near letter quality) printers may be acceptable, and the text can often be darkened by a single round of photocopying which may also blur the dots into a more continuous representation of text. Some experimentation is necessary, and justified.

Low- and high-resolution fonts, magnified.

Summaries for primary research papers are often typeset and there is less pressure to produce perfect copy. However, the summary should still be passed to a non-specialist colleague for comment and the typeset version should be carefully read and checked; it forms the first line of communication between the authors of a paper and the reader. Similar arguments apply to your thesis summary.

Progress reports

It would be surprising if your department did not require that you write a fairly detailed literature survey and progress report. This report will usually contribute to your assessment by a Ph.D. Committee and the relevant funding agency. It will also help you take stock of your work and to plan future directions.

A progress report will not normally be unduly long. Local requirements will vary, but as a rough guide you would aim to produce between 20 and 30 sides of double-spaced A4 text, excluding Figures, Tables and references. Typically, you might expect to invest between one and two weeks to produce this report. Of course, this need not mean two continuous weeks away from the bench; as a professional player you will have discovered the deadlines well in advance, and will have started early.

Most progress reports will require that you describe the work done to date, as well as your future plans. Thus, it contains the elements of the two major forms of written communication used by scientists: the published paper, which is largely retrospective, and the grant application, which is mostly prospective. Your report will be a hybrid of the two, and your Department will advise you about the balance that you are expected to adopt.

Writing a thesis

Your thesis will stand as your personal record of your achievements over the period of your research project. It is the document that is submitted to your examiners, and which will earn you your higher degree. You might be a co-author on research papers as well, but as the term 'co-author' implies, you will not have assumed sole responsibility for the writing of the papers. When you come to write your thesis, you must recognize that it is perhaps five to ten times as long as a typical research paper and will be written, virtually unaided, by you. Decisions about content and structure should be made in consultation with your supervisor, and in an ideal world, these discussions will take place long before you have to start writing. Your experimental work must always be done with due consideration to your thesis. Your personal bibliography should be a resource to be used in your thesis as well as a research aid. Literature surveys should find their way into your introductory material. It is never too soon to start planning for your thesis.

Writing a thesis is much the same as writing other scientific texts, and the advice given in the previous sections will apply here also. Like research papers, it will contain a body of text, and numerous illustrations that must be consistent in style, and show evidence of logical thought. Yet in other respects, a thesis is rather different. It is much larger in scope, structure and production. A 'typical' higher-degree thesis is hard to define, but might contain 150–200 pages, 30 000– 60 000 words and maybe as many as 100 Figures and Tables. These figures are not given as guidelines (theses are too varied for such advice), but to illustrate that this is a major undertaking and will not be completed in a couple of months of sporadic evening work. A more accurate estimate might suggest at least three months of uninterrupted writing, and at the time when you have reached the peak of your experimental skills. It follows that the earlier you make a start, the less pressure there will be on you, at this most productive time, to abandon your laboratory work and start writing. In the first part of your final year you could devote one night a week to thesis writing, expanding to one, then to two full days a week and gradually working up to full-time writing.

Local regulations

You will be bound by regulations on thesis preparation in the ordinances of the institute that has registered you for your higher degree. However, detailed advice and information will often be lacking, because such advice is very subject-specific. Your department may have further 'guidelines' that are either compulsory or strongly recommended. You must meet both these sets of recommendations, and should only depart from them if you are given permission to do so. There may be one exception — some ordinances will assume that you will prepare your thesis on a 'steam' typewriter and will not acknowledge such 'new-fangled' ideas as word processing, desktop publishing and computer-generated graphics. You may wish to re-interpret these ordinances with a modern perspective, but once again, seek advice. Even if it is possible, printing your whole thesis in 8-point type (the size of the running footer on this page) will not endear you to your examiners, and you may have to reprint the whole thing. Here are a few suggestions on thesis structure and content that apply to the biological sciences, but make sure that if you decide to use any of them, it is permitted by your local regulations.

Overall arrangement of chapters

There are several ways to structure a thesis but only two are commonly employed. The format of the thesis can follow that of a typical paper, and comprise sections entitled Introduction, Materials and Methods, Results and Discussion. The Results section can be subdivided into several chapters or divisions. This format is often of value when very similar methods have been used extensively throughout the project.

Alternatively, the study might have comprised a number of areas that used essentially non-overlapping methodology. In this instance, a common format starts with a General Introduction, followed by several chapters, each consisting of Materials and Methods, Results and Discussion. Finally, the whole thesis is made coherent by a General Discussion. Detailed methods of data analysis or computer programs can often be usefully included in an Appendix or Appendices.

In addition to the chapters of 'substance' in a thesis, there are many additional sections that must be included. All but two of these are not optional. They are, usually at the front: the abstract, contents, list of Figures, list of Tables, list of abbreviations, acknowledgments and optionally, a dedication. At the end of the thesis are placed the references and, optionally, a list or collation of papers that have been published during the research project. The stiff board covers of some theses incorporate a pocket in the rear cover for these papers.

Specific items within a thesis

Although individual theses will vary considerably, there are certain elements that should always be included, even if not specified by your local requirements. Consider including most or all of the following, even if you are not obliged to do so.

Abstract. This should be a concise statement of the goals and results of the study. It does not usually occupy more than a single page of single-spaced A4 paper (about 450 words). Include your name and the title of your thesis as a heading. One copy is bound within the thesis, and a second is often provided loose for dissertation abstracting services.

Table of Contents. The contents pages are designed to help the reader locate a particular part of your thesis quickly. Try therefore to give as much information as possible. References to the beginning page of each chapter are not as valuable as references to headed subdivisions within a chapter.

Lists of Figures and Tables. Although their value may not be appreciated, lists of Figures and Tables are very helpful to future users of the thesis; no-one else will know your thesis as well as you. A list of displayed items, including their reference, title and page number is required. For example, the third Figure in chapter 2, facing page 76 could be included as 'Fig. 2.3....Title...facing p. 76' in the list. It is a matter of personal preference whether displayed items are placed on numbered or unnumbered pages; if the former, then the word 'facing' is not needed.

List of abbreviations. It is entirely reasonable to use accepted and standard abbreviations. International committees serve to provide unambiguous abbreviations that form part of an international shorthand — consider for a moment the chaos that would ensue if every author used their own scheme of one-letter coding for amino acids. Journals may also have lists of abbreviations that are in common use — check the 'Instructions for Authors'. Any abbreviations that you use must be defined in a list at the front of the thesis. Non-standard abbreviations should be used sparingly, and only if they save substantial space. These must certainly be defined.

References. The list of references or bibliography affords great difficulty to many postgraduate students. Be aware that the majority of examiners check the list for correctness, completeness of the citations, and to ensure that any citations mentioned in the text are included in the reference list. A poorly-constructed reference list gives a bad impression and implies carelessness; some examiners adopt the attitude that if a student only cites the first page of paper they might never have seen, let alone read it! All journal names have an accepted abbreviation (e.g. *J. Mol. Biol.*) — do not construct new ones!

Good bibliographic software will allow the renumbering of a reference list if a new citation is added.

There are two schemes that are used commonly for cross–referencing citations in the text and the full reference. In the numbering scheme, citations are each given a unique number, usually in square brackets [ ] and numbered in the order in which they appear in the text. Although automatic bibliography generation makes this system more accessible, there is little to commend it. It is almost impossible to add another citation and correctly renumber all successive ones. The numbers give no information at all about the work, and it is too easy to forget to use the earlier number when you need to refer to a paper again, later in the thesis.

The most common alternative is the Harvard system. Within the text, citations consist of author(s), plus the year in parentheses (). There are specific rules for one, two and more than two authors (Bloggs, 1992; Bloggs & Smith, 1992; Bloggs *et al.*, 1992) The '*et al.*' is an abbreviation for '*et alia*' (*and others*, in the plural) and thus, cannot be used for a two-author publication.

The format for presentation of a bibliography at the end of the thesis should be consistent. Perhaps you should adopt the guidelines contained within a representative journal for your specialist area, in consultation with your supervisor. Pay particular attention to single- and multi-author monographs. Titles of articles are optional and in the past have rarely been included, probably because of the extra typing and space requirements. Nowadays, a combination of a bibliographic package and word processor should eliminate any extra work, and space can be saved by printing the reference list in a smaller font. For future readers of your thesis, titles in the bibliography will be very helpful, and you are encouraged to include them.

Text. In the past, the text of your thesis would have been typed, double spaced at 10, or preferably 12 characters per inch. Nowadays, you are likely to have access to a laser or inkjet printer that is capable of producing type with much higher quality and flexibility. A dot–matrix printer at 'draft' resolution is faint, does not photocopy well, forms letters poorly and must not be used except for early drafts. At 'NLQ' resolution it may just be acceptable, but establish this before you commit yourself to a particular printer. If you will use a daisywheel printer or typewriter use a new nylon or carbon film ribbon for preparation of the final version.

If you plan to use a laser or inkjet printer, you probably have access to a plethora of fonts (typefaces) at different sizes and in different styles. Resist the temptation to produce a typographer's nightmare, and restrict yourself to one or maybe two fonts, and only a few sizes. Boldface and italic styles should be enough for emphasis, but use both sparingly. Avoid underlining, a method of emphasis that is inherited from the days of typewriters. There really is no justification for outlined, drop-shadowed fonts — they obscure more than they highlight!

Abl
Abl

Serif and sans-serif fonts.

A modern printer will also let you print in vanishingly small font sizes — resist! Use 10 or 12 point, depending on the font that you use. Remember that serif fonts convey more information and are easier to read than sans-serif fonts.

Most word processors allow text to be 'full justified' which means that the text aligns along left and right margins to give an even block of typescript. This can cause 'rivers' of 'white space' to appear in the text, particularly if the word length is high (as in a scientific presentation). Many people prefer to read text that is left justified, sometimes called 'unjustified' (typographically speaking only!).

Many biological terms are single words that can be over a line long. This confuses many word processors. It usually helps to add hyphens or spaces at convenient places within the words to provide sensible 'line break' positions. A good word processor will make sensible attempts at hyphenation, but avoid word-splitting hyphens altogether if you can.

Something else to consider. An older, typewritten thesis would have about 250 words on every page. The density of the same print area, filled with 12 point Times Roman will be about 400–500 words per page. So why are modern theses the same number of pages long as they have been for the past twenty years? If you gauge the size of your thesis by thickness on a shelf you will ignore the recent revolution in printing technology; if your thesis is the same thickness as another from 1963, it will contain more work, and will take longer to read! Is that what you want?

Figures and Tables. Instructions for preparation of Figures and Tables are given later and apply equally to thesis material. Note that it is very unusual for data to be reproduced both as a Figure and a Table; this is unnecessary padding. If photographic plates are incorporated into the thesis they rarely photocopy well enough for subsequent copies and an original print should be included in all copies of the thesis.

Interleaving sheets. This simple, cheap, yet effective device will be a major help to you, your examiners, and those who will subsequently read your thesis. Place a sheet of blank coloured paper between each chapter or major division of your thesis. Each chapter can be located easily, because the coloured sheets are visible when the thesis is viewed from the side. If you want to be showy, print the chapter title or the title and contents on the coloured sheet.

Effective poster presentations

The poster session is now well established as a means of communicating research at a scientific meeting and has largely superseded the short oral communication. Many authors present their work simultaneously, in a single room, by displaying posters upon which they present their data and interpretations. These poster presentations have several advantages. Posters are usually left on display for at least a day, and a specific time is set aside when authors are expected to stand beside their poster to discuss their work. More posters can be presented in the time and space that are available. A timetabled poster session is not subject to the same rigid time constraints as a series of oral presentations and there is no stigma of 'first' or 'last' poster, which, when presented as oral communications would have had unpredictable audiences.

Poster sessions also have their disadvantages. There have been meetings that timetable no specific period for poster discussion, yet have tens or hundreds of posters on display. Other meetings have located the posters in a room so far removed from the interval coffee/trade exhibition that very few attendees bother to go and look at them. If you have to man your poster, how can you discuss other posters with their authors, who are also trapped besides their work? Most of these disadvantages could be dispelled with a little more concern from meetings organizers, and perhaps a realization that the first authors on many posters are graduate students or postdoctoral fellows who deserve rather more encouragement.

If you have attended any poster sessions you will be aware of the range of posters that are presented, from hastily scribbled presentations on the back of a piece of wallpaper to multicoloured, professionally lettered works of art (this format is particularly useful if the poster is intended for display at a series of meetings!). Don't be distracted by the artwork though; a hand-written poster may convey information more efficiently than a detailed, but ultimately poor, graphic design. As in all visual material, planning and design are distinct from implementation and the latter is of less importance than the former. There is a good chance that you will present at least one poster outside your department in addition to any internal presentations; this section contains some information that you may find valuable in preparing this material. Another useful source of ideas is provided by posters themselves; next time that you attend a poster session use the opportunity to look at, and to assess individual posters for their effectiveness as communication devices.

Plan your poster well in advance

In most circumstances, you will have plenty of notice of the need to prepare a poster. Use this time to decide what material you wish to present, what illustrations are needed to convey that information, and whether you need to do 'just one more control in an experiment'. Make sure that you know how much space will be available for the poster to be mounted; the sight of two adults fighting over a 10 cm demilitarized zone between two posters is not endearing. Width and height are both important. In general, you will be given an area that is

Poster formats

about 1–2 m wide and up to 3 m high — discard the lower 2 m unless you are a contortionist! Then, work to an initial sketch that is in proportion to the allowed and chosen area. A typical rough sketch might look like the examples alongside. Experiment with different styles of presentation; can histograms be used in preference to Tables? Can colour help to discriminate information? What needs to be expressed in words? What headlines might be used? You are well advised to discuss your ideas with someone else at this stage, when changes can be made readily.

Three formats for poster presentations are in common use. The first consists of all material pre-mounted on one or two large (0.5 m x 1 m) cards that are transported to the site and mounted in the display space very easily. Only a few drawing pins or pieces of Velcro tape are needed (these items are a legal currency at most poster sessions). The artwork and layout are all decided and prepared long before the meeting. The background card may be coloured and, with the text and data printed on panels of a contrasting colour, can look very impressive and be eye-catching. Disadvantages include difficulties of carrying the poster on public transport, and the weight of the card, such that it might fall off a weak poster board.

The second format is similar to the first, except that the material is all assembled on a rather more flexible sheet of cartridge paper (coloured if required) that can subsequently be rolled up and transported in a cardboard tube. The adhesive used to attach the items must be resistant to 'cracking' when the poster is rolled and it may be necessary to uncurl the poster before mounting.

The third format consists of individual items each about A4 or Quarto sized that are transported in a flat pack and assembled in the correct relative positions when the poster is mounted. Although easy to transport this format will involve a lot of pins, Velcro or adhesive. Some poster boards seem to be specially selected for their impenetrability by drawing pins and refractiveness to adhesive or 'Blu-tak' — the highlight of many a poster session can be watching a distinguished professor removing a shoe to hammer in drawing pins!. This type of presentation will also be vulnerable to the state of the background that is provided by the poster board (complete with graffiti, paint splashes, etc.). It may be difficult to align each individual item so that horizontals or verticals are maintained.

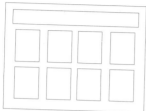

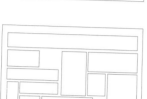

Which design is more interesting? Which is the easiest to follow? Can these two conflicting needs be reconciled?

For the rough layout, enlarge your best initial sketch, keeping the correct proportions. Decide on the size of individual items and try to keep them within Quarto/A4 dimensions as this is the easiest size of printer paper, photographic plate and photocopier paper to handle, although some software for graphics or desktop publishing programs allow 'tiling', such that the printed sheets can be cropped and joined to make a larger composite image.

Roughly sketch each panel at the approximate size and place in position to examine balance and flow of information. Ask your colleagues for comments; you are still at an experimental stage. Look at the balance between illustrations and text, information areas and background space. A rough rule of thumb might be to aim for 50% of the poster area allocated to data and, in turn, at least 50% of the data area devoted to illustrations. Movement of the eye over the poster should be natural, down columns or along rows (see illustration). Size implies emphasis. Try to maintain consistent, invisible guidelines down and across the poster area and make individual panels 'adhere' to these guidelines (identify the guidelines in the example shown here). When you are happy with this stage prepare the individual items as final artwork.

There is a lot of scope for different designs of posters. However, the emphasis must be on simplicity and immediacy of the message. Potential readers of your poster will be walking past at a distance of 1–2 m and text that is illegible at this distance will be ignored. A good ploy might be to use larger text for important conclusions or statements, below which are provided more details in a smaller size of text.

The heading should contain the poster or abstract number, the title of the poster, the names of the authors and their affiliation — an attractive touch is to include your institution logo or crest on the poster heading. Some authors prefer the heading to be mounted on a piece of card, A4 sized, others prefer a banner running across the top of the poster.

9point

18point

36point

Text has to large enough to be read easily. Titles should be bold and between 1 and 3 cm high — the type obtained from the lettering machine at about 36 point would be approximately correct. Where possible, use mixed upper and lower case; words consisting solely of capitals are much more difficult to read. If you use a stencil to add lettering, fill in the stencil outlines to a solid black. Use a word processor that will allow you to preview the document as it will be printed, and use a high-quality printer that can generate the large fonts easily, in a good strong black and lacking 'jaggies'. Try to keep text to a minimum, the reader can always approach you for more details.

Graphs and histograms should aid easy assimilation of information. Rather than use a legend that contains a key to symbols, identify each set of data on the graph itself. Use bolder lines than you might normally use for a paper. A meaningful title is more helpful than 'Fig. 1' or 'Fig. 2', which require that the reader look for reference to the Figure in the text. This is a good opportunity to use assertive titles — 'Transcription of X is induced by Y' rather than 'The effect of Y on transcription of X'. Try to make each panel stand alone, saving the reader from a lot of cross-referencing. The harder you make the reader work to follow the structure of your poster, the less likely you are to sustain their interest in the science.

Other graphic devices can be useful. Arrows, pointing hands or numbered panels all help to indicate the flow of material. Colour can be effective if used sparingly and consistently, for example such that tests and controls are always illustrated by the same colours. Solid blocks of colour can be obtained by cutting up coloured paper and sticking them in position, or by good quality fibre-tip pens. Fluorescent highlight pens have a chisel tip that is good for shading and give eye-catching colours when freshly applied. If you have access to a high-quality colour printer, take even more care over the use of colour; the temptation to produce a garish, but ultimately less readable, poster is a strong one!

At this stage lay out the whole poster in the format that it will occupy at the meeting, on the backing material, if any, that you have chosen to use. If you are satisfied with the overall layout, the poster can now be assembled and packed. Stand the poster vertically and view it from a natural distance. Is it readable? Is the logical flow of your ideas reflected in the construction of your poster?

At the meeting Make sure that the poster is in position as soon as it can be mounted. If the space is already occupied by a poster from a previous session, consider taking it down. Have you remembered the drawing pins? Have plenty of spares because Professor X will be purloining some of yours the moment you leave your poster! Make sure that you are at the poster at the stipulated time to discuss your work.

A good method of disseminating your work is to prepare a photographically reduced version of your poster, fitting on to a single side of A4 paper and to place these in a wallet, attached to the poster and labelled with a 'Please help yourself' notice. However, the reduction can be difficult and may not always be possible. Also, you may find that you have less visitors to your poster if you provide this information.

Taking the initiative

A sad sight at many poster sessions is that of a graduate student standing hopefully beside their poster, watching the world pass them by. If someone comes and looks at the poster, the student waits to be asked about it, and usually, never is. This is far too mild-mannered! At any opportunity, invite people to look at your work, talk about it with enthusiasm, and if someone comes to take a casual look at your poster, consider them trapped, and give them the full story! Treat your poster as a visual aid for your effective, but brief oral presentation of your work. Likewise, when you are freed from your presentation, why not take an interest in the other posters? They will mostly be the work of graduate students or postdoctoral fellows like yourself; just as nervous, equally desperate to discuss their work, but perhaps less confident than you. Help each other, find out about the science in other laboratories and make new friends.

Visual representation of data

We rarely need to communicate the exact details and every datum in our experiments. Rather, we seek to communicate our qualitative interpretation of data sets or a summary of the quantitative properties of the data. Such summaries can be textual; statements such as '87% of patients survived the treatment' convey an image with a degree of precision that is adequate for some purposes. Descriptive statistics, such as 'mean $\pm$ S.E.M. $= 25.6 \pm 1.5$ ($n=25$)' are summaries from which we can construct a mental picture of the data set, if not the exact values (although we must make some assumptions about the data set).

Graphical representation of numeric data allows us to create abstract pictures that, when properly interpreted, will succeed in conveying an image of the data. However, these graphics use a pictorial language, and unless both originator and recipient of the image are using the same language, complete with all of its subtleties, there is an opportunity for misinterpretation or a complete failure to communicate.

Other types of data are better communicated as their original pictorial form. There are no precise textual or numerical descriptions that adequately convey as much information as a northern blot or a gold-immunostained electron micrograph. With graphic images such as these, we receive a huge amount of data but, with practice, can filter out the irrelevant material, and concentrate on the parts of the image that are important.

We also use graphics to construct models of complex systems, and as a form of shorthand. Any textbook in the biological sciences, for example, is full of artistic impressions of the shape, structure and organization of objects and systems, and we take these impressions for granted. A plasmid becomes a circle, a protein molecule becomes a blob attached to a stylized membrane. These types of graphics are not really intended to convey data (the plasmid sequence or the protein structure), but nonetheless require that the viewer adopt the same conventions as the artist. Such representations require careful design and production to ensure that the meaning of the data is not obscured. We interpret most graphics according to experience and unspoken conventions, and when such conventions are violated, misinterpretation is possible.

Computer-based generation of graphics

Your skills as a technical artist will probably be limited. Even if you can draw and use lettering stencils with skill, production of graphics is time consuming. If at all possible, use a computer-based system to generate all graphics. Data graphics packages are discussed in Chapter 6, but a few points of general guidance bear repetition. Make sure that you have access to a high-quality printer, at least 300 dots per inch. Use object-based drawing packages rather than painting packages — the latter can only print out at the same resolution as the screen which is inadequate for presentation graphics. Avoid the temptation to use lots of fancy 'effects' such as three-dimensional scaling — these reflect the 'business' origins of much of the software and (as seems to be the intention with some business presentations!) impede interpretation of the data. Use a 'mainstream' package and save graphic files in a format that can be read by other programs; this reduces the danger of being locked into a program that becomes obsolete.

See Chapter 6, 'Computers and computing'

Scattergraphs and line graphs

In the biological sciences, graphs are mostly used to plot two series of data with the independent variable on the abscissa (*x* values) and the dependent variable on the ordinate (*y* values). Several series of *x,y* data are often plotted on the same graph, discriminated by differing line types or symbols. The graph should not be used to represent discontinuous quantities, as the reader assumes an even distribution of data along both axes.

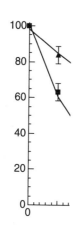

Axes. Axes should extend to the nearest whole number, easily divisible, that exceeds the largest data point. In many instances, and particularly where a set of data are plotted logarithmically, a zero point is either unnecessary or impossible. Under these circumstances, the axes should not be joined. If outlying data are to be plotted, it is permissible to break the axis, but indicate the break clearly.

Tic marks. These should be few and should not clutter the graph. Current conventions usually demand that the tic marks (less than 5% of the height of the other axis) are placed inside the plotting area. Major (large) and minor (small) tics can be used to subdivide an axis further.

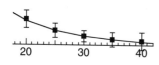

Tic labels. These should not use too many digits and should, if at all possible, avoid exponential notation. The tic labels can be simplified by appropriate divisors or multiplicands in the data and even better, by making use of prefix notation in the axis label (nmol instead of mol x 10^{-9}). If the axis is plotted on a logarithmic scale, choose evenly spaced tic labels that hint at the logarithmic base used (usually base 10 i.e. 0.1, 1, 10, 1000...or base 2, i.e. 2, 4, 8, 16...).

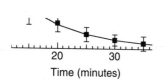

Axis labels. Axis labels should be informative, in terms of both the parameter displayed and the units that are used to express the magnitude of that parameter. The correct format for an axis label is: Parameter (units of measurement). An axis label such as 'units of enzyme activity x 10^{-4}' or worse still 'enzyme activity (colorimeter reading)' is much less informative than 'Ornithine decarboxylase activity (nkat/mg protein)'. Where possible, use S.I. units, these are readily manipulated by prefixes.

Symbols. Symbols used to identify data points should be of the restricted set commonly found in journals and which are readily available as dry-transfer symbols. Avoid using the more esoteric symbols that are often found on the sheets, and in particular, those symbols that look like US aircraft decals! Squares and circles are easy to centre over the datum, triangles less so. The area in the centre of 'open' symbols should be clear, which means that lines drawn through the data must be added after the symbols. Do not use the same symbol for two different data sets on the same graph. Moreover, adopt an internal logic and consistency in the selection of symbols. The same symbol might be used for control data all the time, another set for a particular series of experiments. Filled and open symbols are often used to represent the presence and absence of a particular treatment. Such cues are tremendously helpful to the reader.

Lines. Lines through the data are a tricky subject. In principle, the only lines that can be drawn over a set of data are the theoretically calculated lines corresponding to the experimental model or hypothesis, as defined by the parameters that are specified. In the simplest example, a straight line drawn through a set of calibration data should have a slope and intercept equal to the 'best-fit' values of those parameters. A line drawn by eye cannot guarantee to have those parameters. The situation is even more difficult for non-linear relationships, where a smooth curve is often drawn on a set of data to indicate a trend rather than a precise mathematical function. This type of line is an interpretive device, indicating the presenter's opinions about the overall qualities of the relationship; whether it falls off at high values, whether it passes through the origin and so forth.

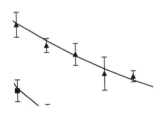

Data may be joined by straight lines, but even this practice makes assumptions about the behaviour of the data between two data points. Joined symbols are often used for elution profiles in chromatography. Some data are included on the graph, but have a secondary role, such as the development of a gradient in chromatography. For this data, individual symbols can profitably be omitted and the data represented as a continuous or broken line. If outliers are plotted, you should calculate the true angle at which the line would join the symbol.

Legends. A 'legend' is used in two different ways. It is most often used to refer to the caption that is placed under the Figure and which contains the definitions of symbols and experimental details and explanation. A legend (or 'key') can also refer to labelling that is placed directly on the graph and which defines the symbols. However, if ambiguities can be avoided, each line can be defined individually by text on the graph. In either instance, the reader is spared from the need

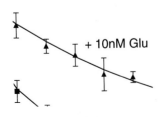

+ 10nM Glu

for repeated reference to the caption to follow the graph. Try therefore to include a key on the graph itself. Note that some journals discourage the insertion of keys on an illustration. This attitude is hard to defend, given the quality of modern graphics software, and provided that the graph is clear and uncluttered, submit a completely annotated Figure.

Error bars. Error bars are used to indicate the variability in individual estimations at a single x- or y-value without including all of the data. The error bars are usually lines extending from -1 S.D. (or S.E.M.) to $+1$ S.D. (or S.E.M.), centred on the datum. It is acceptable to use a single line (from the datum to either $+1$ or -1 S.D.) if it avoids confusion between two lines.

Text. Additional labelling on the graph itself is useful, provided that it is clear and does not interfere with the interpretation of the data. The point of injection of material onto a h.p.l.c. column, V_e and V_t arrows on a gel permeation elution profile, or a chemical structure on a mass spectrum can simplify interpretation of the data.

Histograms and bar charts

Histograms or bar charts use columns, the height of which represent a quantity. Histograms are used to divide up a continuous parameter into discrete categories, exemplified by a size distribution or elution profile. Bar charts are used to compare samples that do not have a continuous relationship, such as in the comparison of a test and control group. Most journals will reject bar charts in favour of tabulated data, as the latter is considerably less demanding of space. In lectures, reports and theses the same restriction never applies and there can be little doubt that the histogram is far superior in the immediacy of presentation of the data.

More complex histograms and bar charts use area to indicate a parameter value; the most common example is the relative specific activity/percentage total cellular protein plot used to display subcellular fractionation data. Care is needed in the use of area to represent a parameter; we are not skilled at comparisons of area whereas linear properties such as height are readily assimilable.

Block width is very dependent upon the nature of the data that is to be presented. In a bar chart, individual data sets are best represented by columns that are separated by a space that should be at least 50% of the width of the column. In this type of histogram, each column requires a label and units are rarely plotted on the x-axis.

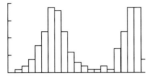

By contrast, a set of data representing a time-series or an elution profile from a column might be best represented by a histogram in which the columns are stacked adjacently with no intervening space. In this format, the serial nature of the columns is emphasized by the labelling along the axis; columns are no longer labelled individually.

Shading of columns is necessary to discriminate groups of data when multiple series are plotted in the same axis area. However, histograms using multiple sets of columns are often difficult to interpret and you should consider the possibility of separating the individual sets into different histograms. If shading is inevitable, then avoid shading or cross-hatching by hand; it always looks uneven. Blocks of adhesive dry transfer shade can be cut to shape and used to fill in a column. Many graph- and histogram-drawing packages include the facility to specify a shade type. Use 'calm' shading such as flat greys, avoiding strong diagonals or hatching that produce moiré effects and the illusion of movement in the columns.

Under no circumstances should you use the 'three-dimensional' effects that are so common in business graphics. These effects may look dramatic, but they impede interpretation of the data, the last thing you should want.

Error bars should extend from the top of the column, upwards for +1 S.D. or +1 S.E.M. Remember to indicate which value is being displayed in the key or legend.

Keys to shading are rarely needed and interpretation of the histogram is improved if the individual columns can be labelled with a meaningful short caption.

Tables

The Table is used for several reasons. First, it is often quite easy to prepare and does not require any graphical ability. Secondly, it is excellent as a device for transmitting the true values of data at high density, contrasting with graphs and histograms that can only give approximate values. Thirdly, it is economical in the space that is used to transmit the data. However, if you prepare a Table that has a large number of data items (about 20) you should investigate alternative methods of presenting the data, either as a set of Tables or as a set of histograms.

Treatment	Groups 1		Groups 2	
	A	B	C	D
Set 1	12±3	6±4	12±4	18±5
Set 2	8±3	9±2	14±4	17±4

123.6
0.004
2.45
12.002

A decimal 'tab' or tabulation facility aligns the decimal point in a column of numbers.

Text in a Table should be kept to a minimum. Never allow the material in one column to encroach on the area occupied by another column. If you prepare a Table on a word processor, use a program that has intelligent tabulation. This means that when the tabulation marker is moved, all items that were tabbed to it move as well. Refinement of the Table then becomes the simple task of moving the tab positions. If the word processor has a decimal tab, use it. This type of tab stop will ensure that numbers align, one above the other, with the decimal points in the correct place.

Headings above columns of data are absolutely critical for easy interpretation of a Table. They should describe the parameter and its units in a compact form. Several lines may be needed for the column headings. Horizontal rules ('straddles' or 'spanners') illustrate the scope of a heading and permit hierarchies of headings to be displayed. Vertical rules, breaking the Table into discrete boxes, rarely improve readability and should be avoided.

Gels and autoradiograms

Many published figures consist of pictures of gels, chromatography plates or autoradiograms. These are often among the worst of all graphics materials to prepare because there are no clear guidelines for their display.

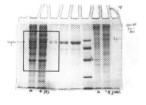

Prints should be of high contrast, but have sufficient tone to retain minor bands or spots. Glossy prints are better. If multiple copies of an illustration are needed (for a thesis for example) then each copy should contain an original print, not a photocopy. Although a photocopier may be reasonably capable of the reproduction of half-tone images there is always some degradation of image quality, such that important information can be lost.

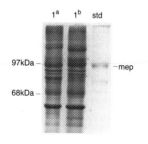

Crop prints of gels to focus on lanes of interest.

As supplied from the photographer, prints contain two types of extraneous material. The first is a white border around the print. Secondly, much of the image itself will be superfluous, and should be eliminated. Unused lanes on a gel, stacking gel regions, unused lower parts and edges of the gel or plate are all distracting and can be trimmed from the print such that only the essential information remains. However, if the image is cropped in this way you may need to give additional information to orient the observer, because many of the visual cues (e.g. the top and bottom of the gel, the full range of markers) have been eliminated.

Labelling of plates requires care. Information on individual lanes, mobility of standards and their characteristic of interest (size, R_f value) must all be added to ease the interpretation of the gel. It is far preferable to annotate individual lanes on the gel with a label that indicates the contents of the lane, than to annotate them as 'A,B,C...' or '1,2,3...' and then produce a complex legend to the Figure. One good way to achieve this is to trim the plate to final size, measure it and draw a rectangle of the required size in faint pencil on a piece of paper. The labelling can then be added to the paper, reproduced as often as required and the prints subsequently attached to the surrounding labels.

Electron or light micrographs should be presented on glossy photographic paper — a photocopy will definitely not do. Include, directly on the print, in white or black depending on which gives the most contrast, a scale bar. When such prints are reproduced in a journal, they will usually be resized to suit the journal format, and a magnification factor incorporated into the legend will become valueless.

Sequence data

Guidelines for presentation of nucleic acid or protein sequences can be obtained from journals that specialize in this research area, such as *Nucleic Acids Research* or *EMBO Journal*. In general, sequences tend to be typed in lines of 60 nucleotides, blocked into six groups of 10. Dot-matrix output is rarely suitable for presentation of this type of material. A daisywheel printer or laser printer with a monospaced font would be optimal. For aligned nucleic acid or protein sequences, proportionally spaced fonts are hopeless, because each character has a different width, making alignment of upper and lower strands or coding regions and translation products rather difficult. Underlining or 'boxing in' of regions of the sequence can be used to highlight important parts of a sequence. Few word processing packages can extend to such enhancements, and it may be necessary to transfer the aligned sequences to a drawing package for embellishment. In this case, since the aligned sequences probably came from another program, why bother with the word processor at all — transfer the sequence data directly to the drawing package. This assumes that the different packages can read each others files, an important consideration in the use of such software (see Chapter 6).

AAAACGTAAA TTGC
TTTTGCATTT AACG

AAAACGTAAA TTGC
TTTTGCATTT AACG

Proportionally spaced fonts (top) make alignment of sequences difficult. Use a monospaced font such as Courier, shown here in the lower example.

Further reading

Booth, V. (1985) 'Communicating in Science: Writing and Speaking', Cambridge University Press, Cambridge

This is essential reading for every scientific writer or speaker. It covers style and grammar with brevity and wit. It may the only book of its kind to advise 'If the typewriter's "e" is full of fluff, please have it brushed out...'.

O'Connor, M. (1991) 'Writing Successfully in Science', Harper Collins, London

This is a useful volume that covers much of the information on scientific writing. The style is easy, and this would make a good reference work.

Reynolds, L. & Simmonds, D. (1982) 'Presentation of Data in Science', Martinus Nijhoff, The Netherlands

This book, and the one below, provide a wealth of information and advice about presentation of data. This book is ideal for those who have limited access to computer-generated graphics. It emphasizes traditional approaches.

Simmonds, D. & Reynolds, L. (1989) 'Computer Presentation of Data in Science', Kluwer Academic Publishers, The Netherlands

A development of the previous volume that emphasizes the use of computers to generate graphics for data presentation. Like the previous volume by these authors, it is full of valuable and practical advice.

Tufte, E.R. (1983) 'The Visual Display of Quantitative Information', Graphics Press, Connecticut

This is a fascinating, witty introduction to the use and abuse of graphics for data presentation. It includes some valuable hints, and highlights common pitfalls. Chapter 4, on the 'data-ink ratio', is a first class illustration of the optimization of graphics to display data at the expense of non-data items.

Talking about your work

Speaking in public

You may as well accept that sooner or later you will have to stand up in front of an audience and talk about your work. This may be at a laboratory meeting, or at a second- or third-year departmental seminar. If your project involves an industrial collaborator you may be asked to give a presentation at a venue outside your own department. If you are very lucky, you will be given the opportunity to speak at a scientific conference. And, unless you are one of the few naturally confident and practised public speakers, you will dread this moment.

Speaking to a scientific audience about your work is just another skill to be acquired. As an undergraduate, you will have enjoyed good lecturers and endured bad ones. You will have listened to visiting speakers that either inspired you or put you to sleep, but can you remember why lectures were good or bad? Could you now turn those experiences to good use, and make yourself a better speaker?

The more experience that you have, the better you will become. Irrespective of how stressful it might seem at the time, presentation of your work or a paper at a laboratory meeting or a journal club is a good introduction to scientific speaking. The audience is non-hostile, knowledgeable and sympathetic. You know everyone and should be able to converse with them just as you do in the laboratory; the only difference is that you are the centre of attention, standing up in front of the group. Enjoy these regular events, and treat them as workshops where you refine your skills. Why not invite specific criticism about your presentation style or your visual aids?

Planning your talk

The recurrent theme reappears: *plan your presentation*. What do you want your audience to remember at the end of your lecture? You should want them to recall the purpose and major findings of your work, your abilities as a speaker and your command of visual aids. You will need to know what you are going to say, and how you are going to say it. Even before you begin, you will know a great deal:

— the subject matter

— the duration of the talk

— the level of expertise of the audience

— the visual aids that are available to you.

Your first experience of public presentation may be a laboratory meeting. These tend to be spontaneous and interrupted; a great deal of planning may not be appropriate. Research group meetings allow you to find your public voice and to learn how to conduct a scientific dialogue. They are less good as exercises in detailed planning. For more structured talks, such as a departmental seminar, it is essential that you are confident about all facets of your presentation. You must plan your talk and supporting notes and design excellent visual aids.

When you start with an outline of your talk you will realize that the subject matter and time are linked. Obviously, your talk should have a beginning, middle section and conclusion, expanded or contracted to fit the allotted time. The relative balance between the different sections will vary with the type of talk that you will give. A specialist audience needs much less introductory material than a general audience.

In your outline, make hierarchical lists of the main points that you want to communicate to your audience. Then consider the background material that you will need to present. What supporting data should be presented? How will you design your visual aids to maximize the communication of this information?

The attention span of any audience is limited, and it is helpful if you break your presentation into discrete sections, particularly if you can produce visual aids that allow a brief summary of each section. These summaries should make natural links to the following material, and indicate your understanding of the logical structure of your work.

You may want to expand your outline into a detailed script. This is valuable as an early planning exercise, but there is the danger that you may be tempted to read from the script, which usually makes for a rather wooden performance. If you learn the text verbatim, you run the risk of forgetting what you have learned. You should aim to be able to talk and construct sentences in 'real time', and if you read or learn a script you will never develop this aspect of your speaking skills. Scientific writing and speaking have sufficiently different structures that a written script can sound over formal. If you must produce a script, say each sentence out loud, and listen to its sound. When you speak, your audience has only one chance to hear and understand, they cannot go back over the material as they can with a scientific paper. Beware of ambiguities such as 'dye-peptides' (dipeptides?).

If you produce a script, condense it back into a series of key points, perhaps laid out on reference cards. Indicate on the cue cards, preferably with a different colour ink, where you will use visual aids. List the key points that you want to make, and use your own words, spontaneously, to make them. Design and use your visual aids so that they provide prompts that remind you to cover key points.

Rehearsals

You will need to practice your presentation several times alone or in front of a few people such as your supervisor or members of your laboratory. Enlist the help of others in your laboratory or department, and have them sit individually at the front, the back, left and right of the room. Try to address each of them in turn, as this will help you to keep your head up and project your voice. Ask them to indicate to you (wordlessly) whether they want you to speak louder, or more slowly. While watching for all this, deliver your talk, and learn how to operate the controls in the room. At each iteration, concentrate on one aspect of your presentation, and have an immediate *post mortem*.

Under no circumstances should you exceed the time allotted for your talk. Indeed, lectures that run a little short are often blessed by an ensuing lively discussion.

Be familiar with the lecture room

If at all possible, practice your talk at least once in the lecture room that you will use for your presentation. Even if you have sat in the lecture or seminar room dozens of times previously, and think you know it well, it is not the same room when you are at the front, looking out on an audience. Make sure that the room is equipped with everything that you need: projectors, chalk/pens, erasers, a glass of water (bring your own if you can). Is there space for you to lay out your notes and overhead transparencies? How do you control the lights and visual aids — if there is a confusion of switches, label the ones that you need (for example, with sticky tape, but remove the tape immediately after your talk).

Giving the presentation

You will undoubtedly be nervous about your forthcoming presentation — this is normal. Even experienced speakers should be a little nervous or tense before they perform; it sharpens the performance and prevents complacency. What can you do to suppress your more extreme feelings of apprehension?

Keep in mind that you probably know more about the subject matter than almost anyone else in the audience. They are willing to learn about your research, and will be sufficiently interested to pay you the compliment of questions at the end. If you are confident about the material, you can concentrate on the presentation.

Slow, deep breaths can do much to alleviate nervousness, and the breathlessness that can accompany it. Avoid alcohol. Often, when you start to speak your voice will take a few seconds to get up to strength. The audience is adjusting to your voice and speech patterns, and is concentrating more on you than what you have to say. So, avoid saying anything of great importance in these first few seconds. One useful device is to run through any acknowledgments at the beginning of your talk, rather than the end. This not only gives you some non-critical material to start, but frees your concluding remarks for a pithy scientific statement, and leaves your audience with a message.

Write this on a 'Post-It' note and stick it where you, but no-one else, can see it easily.

Place in front of you a reference card warning you to speak slowly. Under the stress of the presentation, your adrenaline levels will be high, and you will tend to speed up your presentation. Force yourself to speak deliberately.

Use silence as a communication device. You will tend to want to speak continuously, and will be embarrassed by a few seconds silence as you search for the next transparency. Your audience, on the other hand, will savour those few seconds as an opportunity to reflect on your previous statements. When you display a complex figure, give the audience time to assimilate it, and then guide them through it, a stage at a time. Embellish the data with your interpretations, and then follow up with a diagrammatic representation of the model you are trying to explain. In the old adage — tell the audience what you are going to say, say it, and tell them what you said — but use different words each time!

Do not read the text on your slides or transparencies. This is irritating to an audience, and implies that they cannot read for themselves. Try to use different words in your oral delivery.

Handling interruptions and questions

It is quite possible for the projector to fail, for you to drop a sheaf of notes or transparencies or for you to need a drink of water. Sometimes there will be disturbances from outside. If interruptions like these happen, stop! Wait for the problem to resolve itself and then carry on, preferably backtracking to cover the last point you made. Do not try to talk through the problem; the audience will be distracted by the interruption and will not give you the attention you deserve.

Almost all scientific presentations allow questions from the audience. Usually, these are asked at the end of the presentation, but it is not uncommon for a talk to be interrupted by a member of the audience seeking clarification on a particular point. The latter you treat as an interruption; respond to the questioner quickly and clearly (sometimes with 'I was coming to that next') and then recapitulate your last point before the interruption to re-orient the audience. Questions at the end of the talk can be more wide ranging. You will receive questions from people who want to know how your analysis applies to other systems, questions from people who have failed to understand a point (their fault or yours?), and questions from those who disagree with your data or analysis.

Listen to the question carefully. If it is not clear, ask for it to be repeated. Do not come out with the first thing that rushes into your head. Consider the substance of the question and take time to prepare your answer. Even if the gist of the answer is 'yes' or 'no' you have to say more than that, and explain why you have given that response. You may want to refer to one of your slides or transparencies again — a good reason to make sure they are left in the projector or organized in front of you.

Design and use of visual aids

Your talk will almost inevitably be illustrated by visual aids, and these will usually be 35 mm slides or overhead transparencies (sometimes called foils). If you need any other type of visual aid, such as a video tape you will probably need to consult your local audio-visual technicians for advice. Here we concentrate on the two most common visual aids.

From your own experience you would probably agree that the quality of slides or transparencies presented at a scientific meeting (or in any lecture) is variable and that visual aids that are ineffective greatly reduce the impact of the scientific message. Unless you take the preparation of all visual materials seriously there is a good chance that you will be presenting equally bad material at some time in the future. This is by no means inevitable; consideration of a few simple points can go a long way towards ensuring that your scientific message is not obscured by the slides or that its reception is not impeded by the hostile response of an audience to bad materials. The following guidelines may help you in the preparation of slides and transparencies. Additional advice on the visual display of data can be found in Chapter 4.

It is rare that you will be asked to present a lecture at 1 week's notice; usually, the date of an impending presentation is given several months in advance. There can be little excuse for a last minute panic in which quality and critical appraisal are abandoned in the rush to have the visual aids printed just before the presentation. Give yourself plenty of time to design slides and refuse 'generous' offers such as '*If you like, you can use my slide on that — it's somewhere in my folder — it's a bit faded now and the glass is cracked but you can still read the data and you only have to refer to the first, ninth and thirteenth lines anyway*'. Remember that the person offering you the slide isn't the one giving the lecture!

How many slides or transparencies?

Exact guidance is impossible and unnecessary here. The range is obviously from one slide (on all through the talk and containing all the data!) to as many slides as the projector can display in the allotted time period (about one slide/second!). The acceptable intermediate is variable and depends on the nature of the slide (background information, diagrammatic or data?) and the amount of explanation the slide will require — this will be discovered during rehearsal of the talk and will be influenced by the design of the slide. A rough baseline might be a slide or transparency every 1 to 2 minutes, but considerable variation is acceptable. Some speakers use slides to amplify points, provide conclusions and to lighten the tone of the presentation. Limited use of cartoons can be fine if you are confident about using them to deliver a *relevant* message, but avoid material that might be misconstrued as sexist, racist or otherwise offensive.

The perfect 35 mm slide

There is such a thing as the perfect slide — it's just a very rare species. Here are a few guidelines to help you attain or at least approximate perfection.

- It should have a clear purpose. An effective slide should have a main point and not just be a collection of data. If the central theme of the slide cannot be identified immediately, it would probably benefit from revision. Avoid this by designing the slides in advance.

- It should be readily understood. The main point should be immediately apparent to the audience and be understood quickly. While trying to work out what your slide means, the audience is not hearing a word you say.

- It should be free of non-essential information. Information not directly supporting the main point of the slide and not important enough to be mentioned could probably be held in reserve until question time.

- It should use a graphic format if possible. In graphs and histograms, qualitative relationships are emphasized at the expense of precise numerical values, while in tables the reverse is true. If the main point of the slide is to indicate a qualitative relationship it is probably better to use a graph or a histogram than a table of data — the latter, although much easier to prepare, is also harder to assimilate.

- It should be legible. We've all endured slides that are too small, untidy or too faint to be readable. It is rare for us to complain about slides having too little information on them or text that is too large — maybe that is because the scientific message is getting across almost without us being aware of it.

Good slide design will take into account the fact that the slides will be projected across a large room, not squinted at on a light box. A good set of guidelines has been compiled by FASEB (Federation of American Societies for Experimental Biology) and are worth summarizing here.

For a viewer seated at the rear of an 'average' lecture room, the eye-screen distance is about 10 times greater than the width of the projected image. This ratio is useful in that it permits you to assess by eye the approximate appearance of the slide by holding the slide at a distance of 35 cm (14 inches) from the eye. More surprisingly perhaps, this ratio is also correct when a diagram on A4 paper is held at a distance of 3 m (10 feet) from your eye. It is a good rule of thumb that anything that is not legible at these distances cannot be trusted to project well.

FASEB also calculate that the minimum size of legible lettering (for normal vision) is about 1/60th of the projected image width. For a standard 35 mm slide this transforms to a typing area 42 characters wide by 14 single-spaced lines high (assuming 12 characters/inch). This window is about 9 cm wide by 6 cm high. Typescript that can fill this area should be readily legible when projected.

Slide design and production

A set of slides for a 'typical' presentation can contain a range of materials from different sources. Pictures of gels or blots, photomicrographs, line graphs and diagrams, tabular data and simple textual information must all be assembled into a coherent whole. Try to prepare as many of the slides as you can in a common style and format.

There are two different strategies for production of slides. In the first, you assemble the images on paper, and produce the text and artwork by hand or with the help of computer software. These paper images will then be photographed and turned into slides by standard photographic procedures. This is the most common method of slide production.

Alternatively, images created using specialized computer programs can be processed directly to slides. Although more expensive, this option allows colour slides to be generated at high resolution. Your institution will probably have a central photographic department that can accept disk files of slide images and generate the finished product, but before you embark on this route, make sure that the department can support the hardware and software that you will use. Discuss the cost with your supervisor!

Serifs: Il-1

No serifs: Il-I

How many different ways are there to interpret this term, as printed in the different fonts?

If you have to generate paper images first, remember that a typewriter is not particularly good for generation of slide material. Gain access to a computer system and a high-resolution laser or inkjet printer that allows you to generate text in different fonts and sizes. Bold, sans-serif fonts, such as Helvetica, give a very clear image, but are not as easy to read as fonts with serifs, such as Times Roman. Use wide spacing between text lines.

If you have to use a typewriter, make sure that is equipped with a carbon ribbon — this is superior to a nylon ribbon which produces characters that are fuzzy and look out of focus.

Items other than text should be prepared simply and boldly, perhaps even more so than you would use for a paper publication. Diagrams and graphs should use bold lines and should not include too many tic labels, as these are distracting and often force you to use a smaller size of lettering. Avoid taking an illustration directly from a paper — this will often contain insufficient labelling and require a lot of explanation. Don't expect your audience to be able to read the legend that was printed underneath the figure.

What type of slide? Black-on-white slides are easy to prepare at high contrast, require least preparation of artwork and are easily handled by most projectors. They also give a degree of illumination in a darkened room.

White-on-blue slides look good when new, but fade quickly to a greyish blue that loses all of the initial advantages of high contrast (this fading does allow you to ascertain the age of the data that is being presented!). These slides require special processing.

White-on-black slides are relatively easy to prepare, but require the use of a good high-contrast film to allow the white text to be clearly defined (they can look like light grey text on a dark grey background). They give very little illumination when projected and demand a well-dimmed lecture room; this may prove unfortunate if you need to refer to a script or expect your audience to take notes! They have an advantage in that suitable pens can be used to colour areas of text directly on the slide to highlight important information (but this is risky — have spares made!). Some autofocus projectors may not respond well to these slides.

Full-colour slides prepared from paper masters may look 'washed out'. Their preparation requires additional photographic skills and extra preparation time. In general, they are not recommended. If you want to photograph images from a video monitor make sure that the photographer knows how to do this, and tries several exposures. To start, try 1/15th sec at f5.6. Aim for strong colours, and bright images. The disadvantage of these imagesis that you will not often be able to add much annotation or crop the image. Good knowledge of computer hardware and software can often overcome such limitations.

Colour slides prepared by a film recorder connected directly to a computer look superb, and can generate a huge range of colours. The tendency to use garish or over-fussy colour schemes is great. Before you prepare this type of slide, select and stick to a good but simple colour scheme. Also, these slides are expensive.

Slide maintenance

In Europe, the majority of projectors will accept slides that are glass mounted, i.e. the film is mounted between two thin plates of glass that protect the emulsion from dust and scratches. These glass plates are very thin and fragile so you should store and transport them in a sturdy box or wallet designed for that purpose. It is also a good idea to have a spare mount or two in the box, just in case of accidents.

Most mounted slides have a white face and a dark face. The white face is intended to face the light source (a very hot quartz–halogen lamp) when the slide is being projected and reflect much of the heat that can otherwise, under extreme circumstances such as a failed cooling fan, melt a slide in a dramatic manner. The race to explain data before it slithers off the screen is visually exciting for the audience but less so for the presenter!

In some countries, slides are mounted in thin cardboard carriers that lack glass plates. These are less durable, and can flex and jam a projector. Plastic mounts of similar design are preferable.

Spotting slides

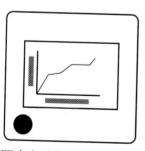

With the slide on the desk in front of you, such that you can read it properly, the spot is placed at the bottom-left corner. The white face of the slide will be facing you.

A major drain on audience sympathy occurs every time a slide is projected upside down and/or back to front. This can largely be avoided by 'spotting' the slides correctly. With the slide on the desk in front of you, in the correct orientation so that you can read it without resorting to yoga, and with the white face upwards, place an adhesive spot at the bottom-left corner of the mount. Number the slide on or near the spot.

If you load the slides yourself, there is no excuse for incorrect projection, and you should become familar with the loading and operation of a 35 mm slide projector. To load the slides into the projector rack or carousel, position yourself (hypothetically) behind the projector and facing the screen. The white face of the slide will be facing you and the back of the room, and the spot should be visible in the upper right corner. If you load the slides with your right hand, the spot should be covered by the thumb of your right hand. The slide contents will only be readable with difficulty, because in this orientation, the image on the slide will be upside down and back to front.

Design and use of overhead transparencies

Although the majority of scientific presentations still use 35 mm slides as the primary medium, the overhead transparency has a number of features that make it useful, especially for internal research group or departmental use.

Advantages

Overhead transparencies have a number of specific advantages. The projector is simple to operate and gives a good image in well-lit rooms. Moreover, the projector is under the complete control of the presenter (no projectionist is needed). It is relatively easy to prepare a wide range of materials (both monochrome and colour) and the ability to add overlays gives a unique opportunity to build up an argument or concept visually and textually at the same time. Materials can be prepared by conventional photographic processes or they can be hand-drawn; the latter is more appropriate for informal presentations such as laboratory meetings. The transparencies can be modified during the presentation. Although you can use a sliding mask to reveal information, there is a school of thought that maintains that this is irritating to the audience.

Disadvantages

An exaggerated keystoned image

Disadvantages of the overhead projector include 'keystoning', in which the beam of light is angled upwards to a vertical display surface, causing a disconcerting distortion and making the image wider at the top than the bottom. In a small room, the projector may be tilted so that it is normal to the display surface, while in a larger room, keystoning is usually avoided by angling the screen towards the projector so that it is normal to the incident light. Not all display screens can be angled in this way. Although the projector is under the complete control of the presenter, this imposes an additional responsibility of ensuring that the image is legible, correctly focused and angled. The transparencies can be difficult to control as they may stick to each other, and a long presentation requires their careful management. Another irritation of overhead projectors is that they contain fans that can be noisy and generate a force 10 gale that wreaks havoc with a sheaf of notes or transparencies!

Many projectors are equipped with a roll of clear acetate film that is used as required, and scrolled forwards to give a new, clear area. Scratching, smudging, the inability to prepare material in advance and the problem of referring back to material are sufficiently persuasive that the acetate roll should not normally be considered for a pre-planned presentation, although it is adequate for an *ad hoc* discussion.

Clear acetate sheets, usually A4/Quarto or slightly larger are readily prepared in advance, will lie flat and can be interleaved with thin sheets of paper to prevent sticking.

Laser printers and photocopiers accept special transparencies

A special type of transparency can be inserted into a photocopier in place of a piece of blank paper; this allows ready transfer of a conventional paper image to a transparency. Make sure that you use the correct transparencies. Standard transparencies will not survive the high temperatures within the photocopier and will damage it. If in doubt about an unmarked transparency, never assume that it is suitable for use in the photocopier.

Transparencies can also be passed through a laser printer, but this type of printer fuses a toner to the paper, in the same way as a photocopier does. Again, make sure that you are using the correct, heat-resistant transparencies for a laser printer.

Inkjet printers can produce high-quality images, and there will be special transparencies for this type of printer. These will have a rough, absorbent surface that will take up a lot of ink without smearing, ensuring a strong image on the screen. This type of printer can also generate colour images at high resolution and with good saturation, but choose bold colour schemes to prevent the image being washed out as it is magnified on screen.

Pen plotters can generate attractive coloured transparencies, but you will need to use special plotter pens and transparencies for a stable, even image. Plotters are not very good if you will need large areas of solid colour.

Simple transparencies are prepared directly by writing on the acetate with special fibre-tipped pens or by preparing the image in advance on a sheet of ordinary bond paper, photocopying it, and colouring in or highlighting important parts afterwards. In general, the second method is preferable. You will have an exact paper copy of the transparency for notes, reminders and guidance during the lecture. The original image can be used to prepare a number of foils, giving an opportunity to rectify mistakes or to emphasize different parts of the same image in successive transparencies. Moreover, 'Tipp-Ex' correction of a white paper image is usually more successful than a finger moistened with saliva, which tends to have effects beyond the immediate correction. Never correct a transparency this way in front of an audience — magnified on screen, it looks revolting!

Never reproduce on transparencies material that was intended to be read at close range. Normal typeset or typewritten text is too small for effective presentation at lecture theatre dimensions, although it may just about be acceptable in a small group meeting (in which case, why not give everyone a photocopy of the material?). A particular disaster is the reproduction of a complete DNA or protein sequence from a computer printout — it will be unreadable, ineffably boring to all but the presenter and will obscure the message that is being transmitted. If you wish to highlight a region of sequence for some reason, then the whole sequence can be represented as a straight line to put the region of interest into context.

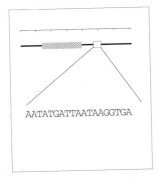

AATATGATTAATAAGGTGA

If you are preparing transparencies by writing on them directly, do not write in your normal handwriting — it will usually be too small and too stylized for effective communication. Resurrect the 'penny and post' writing of your early schooldays and make your text larger than normal. A sheet of ruled paper placed under the transparency will help to keep the text on the horizontal; use every other line as a guide. Two types of special transparency pens are available in a range of tip thicknesses. Ordinary felt-tipped pens rarely work well on the surface of the acetate. Water-soluble pens are useful for ephemeral presentations, after which the sheets can be cleaned by running them under a tap or by wiping them with a damp cloth. These are most useful for laboratory meetings and small group presentations. Permanent pens should be used for any important presentation, particularly if you will be using the transparencies several times during rehearsals. Water-soluble pens are vulnerable to the clammy hands of a nervous presenter and the inadvertent removal of half of the data on a transparency is a disconcerting sight, especially for the presenter! The 'permanent' pens are in fact readily removed with ethanol.

Author's note: I am reliably informed, by a respected colleague, that if you run out of blank transparencies at 11pm and have to complete a lecture for the following morning, gin is an effective solvent for 'permanent' pens!

When you acquire a set of pens please do everybody a great favour. Throw away any orange or yellow pens immediately, use them as highlighting pens on ordinary paper, or give them to a favourite niece or nephew. These colours do not produce a readable image and because they are little used, the pens sit in the wallet until all other colours are used up or dried out, at which point they are brought into service for the whole of every transparency. Be zealous about this — if you see a colleague with an orange pen, throw theirs away as well!

The preferred colours are black, blue, green and, to a lesser extent, red. These ought to be sufficient for any transparency, especially as several studies have demonstrated that use of colour favours the retention of peripheral rather than central information.

Overlays are an effective device for constructing an image slowly, but make sure that they are in the right order. Either include registration marks (small dots outside the main display area) to ensure correct alignment or tape the overlays to the sides of a mount so that they register correctly when folded into position.

Operation of an overhead projector

The presenter has considerable responsibility for the quality of a lecture using an overhead projector (OHP). It is worthwhile spending a few minutes, in advance of the lecture and preferably before the audience is installed, making sure that it is correctly set up.

The projector is normally adjacent to the user, and the projected image is either directly in front of the audience or to one side. Position the projector so that it displays the best image, as free from lateral or vertical distortion as possible.

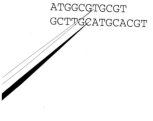

ATGGCGTGCGT
GCTTGCATGCACGT

Focus the projector using a test sheet with some fine lines, preferably derived from a photocopier-produced transparency. Make sure that the lens and glass surfaces are clean and free from grease and dust (especially if there is a blackboard in the room); a soft, lint-free cloth may be used to clean both.

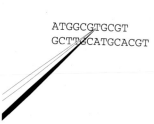

ATGGCGTGCGT
GCTTGCATGCACGT

During the presentation, the transparencies are placed on the surface of the projector, such that they can be read directly by you when facing the audience from a position behind the projector. You might prefer to switch it off when changing transparencies; this is less distracting than displaying the changeover from one transparency to the next. Don't be afraid to switch off the projector when it is not needed; it is competing with you for attention.

When pointing at the screen, make sure that the pointer and shadow don't point at different things — bring the pointer close to the screen.

Items on the transparency can be indicated using a pencil, knitting needle or similar pointer directed to the surface of the transparency. Bring the pointer close to the surface so that the shadow it casts on the screen is correctly focused. This method of pointing does have the disadvantage that tremors in the hands are emphasized; you may prefer to use a large pointer and point to the image on the screen. If you do so, avoid touching the surface of projection screens; they are often fragile and susceptible to scratching.

The *viva voce* examination

Some months after you have completed and submitted your thesis, you will undergo a *viva voce* ('live voice'), or oral examination. This oral examination ('*viva*') is a formal occasion. The examiners will have scrutinized your thesis in some detail and in this examination, which typically lasts from 2 to 3 hours, you will discuss your work with at least two examiners. One of the examiners will come from outside your institution (external examiner), the other usually from your department or institution (internal examiner). Sometimes there may be additional examiners present, and your supervisor(s) are sometimes given the option of attending. It is rare nowadays for the internal examiner to be the supervisor — this practice should be generally discouraged as a little too incestuous.

It is difficult to predict the format and route that the *viva* will take. In all cases, the examiners will seek to establish that you know the background literature, that you performed the experimental work and that you understand the experiments and the conclusions that derive from them. They must be persuaded that the thesis is your own work, and that it and the science it contains are at a standard suitable for the award of a higher degree. If your thesis is scientifically sound and well produced the *viva* may take the form of a broad ranging discussion, in which you are given an opportunity to extend your ideas immediately beyond the substance of your thesis. At the other extreme, the examiners may have doubts about your thesis, and be seeking reassurance, or confirmation of their worries — fortunately, this is a rare event. Between these two extremes we find that the most common *viva* involves a detailed discussion of the thesis, focusing on some of the weak parts, but, with due attention given to the better parts as well.

The more confident you are about your thesis and your work, the less traumatic your *viva* will be. You will usually be consulted about, and given a date for the *viva* with plenty of notice, and thus, you can plan your build-up to this examination. Some weeks before the *viva*, read through your thesis once more, making notes about the main points of each chapter, making a list of typographical and other errors, and reminding yourself of the experimental systems that you used. It is not unknown for research students to use a technique for three years and, at the end of this time, still not really understand the equipment that they are using — the inability to answer a simple question such as 'how does a scintillation counter work?' can only weaken the examiner's opinion of you (and to a lesser extent, it must be said, of your supervisor and your department).

Many *vivas* start with an invitation to the student to summarize the work that they have completed. Make sure that you have produced an outline of the answer that you would give, and be prepared for questions such as 'what is the significance of your research?', 'how would you approach this problem if you had your time over?', 'what skills have you developed during your studies?'. These questions are often intended to allow the student to relax before the detailed scientific discussion starts, yet that can have exactly the opposite effect. You have been so embroiled in your work for so long that it is difficult to stand back from it and answer such questions. Make sure that you consider such issues before the *viva*. You might want to discuss such points with your supervisor beforehand, but make sure that any discussion focuses on your ideas. Explain why you want the discussion, and ask them to bear with you as you expound your ideas.

As the *viva* date approaches, it might be a good idea to check, by a literature search or by perusal of recent journal issues, whether any relevant papers have appeared. There is a chance that the examiners have yet to read them, giving you an opportunity to impress. Have another read through your thesis and think about what you would like to discuss in each section. The more you have to say, the more likely it will be that the *viva* will go in the direction you want

On the day, arrive in plenty of time, bringing with you your copy of the thesis. If you have to travel a good distance for your *viva*, it is advisable to take your notebooks and any floppy disks that contain thesis data or text. Then, if the examiners recommend minor changes it would be far preferable to complete them immediately and re-submit the corrected version straight away. You should dress in a way that indicates you recognize the formality of the examination, but are comfortable.

At the end of the *viva*, you will be asked to leave the room for a short period while the examiners deliberate. Then, you will be informed of the decision. It may be that the decision is not what you expected — if so, listen carefully to what the examiners have to say. Make absolutely sure that you are clear about their recommendations and what you might have to do to bring your thesis up to the required standard. Discuss these requirements with your supervisor, and make the changes as soon as you can.

Further reading

Booth, V. (1985) 'Communicating in Science: Writing and Speaking', Cambridge University Press, Cambridge

Booth's chapter on 'Speaking at scientific meetings' should be compulsory reading for everyone who has to stand in front of an audience.

O'Connor, M. (1991) 'Writing Successfully in Science', Harper Collins, London

This book includes some good material on oral communication in science, although the emphasis is on written material.

Phillips, E.M. & Pugh, D.S. (1987) 'How to Get a Ph.D.', Open University Press, Milton Keynes, U.K.

Although written for students, the overall tone of this book is a little pessimistic, concentrating somewhat on the problems rather than the highlights. It should also be read by supervisors, who sometimes forget just how vulnerable a postgraduate student can feel.

Sindermann, C.J. (1982) 'Winning the Games Scientists Play', Plenum Press, New York

This second volume by the author of 'The Joy of Science' covers, in very readable style, the many skills and activities that define a professional scientist. The chapter on 'The Scientist as Performer' is particularly relevant here.

Computers and computing

Introduction

Almost every laboratory technique has seen the introduction of computer technology for instrument control and data collection/processing. In many cases the data are such that they can be processed more easily by machine than by a human: e.g. an autoradiogram, a macromolecular sequence or structure, an h.p.l.c. run or a database of all known nucleic acid sequences. Digital computers are cheap, and keyboards and VDUs will feature prominently in any reasonably equipped laboratory. To illustrate, in the author's laboratory a gas chromatograph/mass spectrometer, spectrophotometer, h.p.l.c. and laser densitometer are constructed as largely featureless boxes, with few or no user controls except a power switch. Sitting beside each of these machines is a computer that is responsible for all of the instrument control, data capture and analysis. Also in this laboratory are a further three computers used for screening of the literature through CD-ROM databases and *Current Contents on Disk*, production of posters, reports, graphs, papers and theses, maintenance of personal and laboratory-wide literature databases, and nucleic acid and protein sequence analysis — a total of seven machines, a total of over 1500 megabytes of hard disk space and a constant demand on laboratory resources, staff and technician time. Equally importantly, these machines place a demand on the intellectual skills of the people in the laboratory. In return, sophisticated and powerful analyses are straightforward and data can (usually) be moved from machine to machine, or from instrument to paper or thesis with relative ease.

See Chapter 3, 'Conducting a research project'

This is a common scenario in a modern biological laboratory and, to make the most of such resources, you will have to become familiar with computers and computing. With luck, you will have been taught the rudiments of computing in your undergraduate courses. If not, it is essential that you learn basic computing skills at the earliest opportunity. Just as you will become skilled in other techniques, it is reasonable to expect that you acquire a working knowledge of computing, as part of the training that you will take with you when you move on to the next stage of your career.

Different research projects demand different levels of computing skill which should have been clarified before you accepted the position. Your research might require the development of new software tools, demanding that you acquire extensive programming skills. Alternatively, you may need to build a thorough knowledge of existing software, which may itself contain a programming language at a higher level. At the very least, you will have to build a basic understanding of the computer-based representation and storage of data, and the manipulation of files of data on different media. It is not the purpose of this chapter to provide detailed instruction in computing; computer systems and individual needs are too varied. Rather, a few overall comments will serve to illustrate the scope of what is available to you, and indicate paths that you might take. It is not an unreasonable goal to expect that at the end of your research programme, you should have a working knowledge of at least one computer system, be able to enter and edit a file, know how to transfer and back up data files and be fairly familiar with at least a few software packages.

Hardware

Although most of the computing that you might need to do can be conducted on a range of different computers, your choice will be restricted by the laboratory environment and the availability of local facilities. Nonetheless, for the computing activities that are specific to your own research, you may be able to choose between a number of hardware 'platforms'. This choice is controlled by the availability of software to perform the tasks that you need, and the performance of the computer relative to the demands of those tasks. Computers of recent design are faster, and tend to offer more facilities, but, such performance increases are often rather superfluous. Also, paper specifications can be misleading — we rarely need to know how fast a computer can work as an electronic switching device; what is important is the performance of a particular piece of hardware when running the appropriate software. A fourfold reduction in speed for a given software task from 100 ms to 25 ms is only significant if you perform this task repeatedly and automatically. If the fourfold reduction is from a 16 minute to a 4 minute operation, the advantage is clear, even for tasks that you perform infrequently.

For most computing tasks, the machine is idling, and waiting for the user to invoke a command or enter data. After an appropriate command, a burst of computer-intensive activity may ensue before control is returned to the user. Our perception of the performance of a computer system tends to be based on these computationally intensive periods. A faster computer will only really make its enhanced performance apparent at such times, and again, it is necessary to balance performance with need.

A major development in computing in recent years has been the development of user interfaces that facilitate interaction with the software (see below). These interfaces are themselves large programs that need powerful hardware to run effectively. You may need to use high-quality hardware to run a graphical user interface, just to get reasonable performance from a relatively simple program.

You will find that the current generation of personal computers offer more than adequate performance for the great majority of tasks that you wish to conduct. At the other extreme from 'personal' computers are the workstation and mainframe systems, to which you can connect as one of many users, and make use of shared resources. Inevitably, these systems are more complex than single-user systems, and at first it may seem very difficult to get anything done. However, large systems will give you access to very powerful software; your data will usually be backed up automatically, and you can gain access to national and international computer networks.

User interface

Although there are many specific user interfaces, they follow two overall patterns. The first, the 'command line' interface, exemplified by MS-DOS, demands that you tell the software what operation to perform, by typing instructions. This interface is fast and can be very efficient, but suffers from the disadvantage that you have to know the commands before you can do anything. A good on-line help system is mandatory - if in doubt, try typing the word 'help' or a question mark, and then press the return key. The second type of user interface, of which there are many lumped together under the acronym GUI (graphical user interface) is rather different. Interaction with the software is usually through a mouse, selecting menu items that 'pull down' as required; this type of interface is seen in the Macintosh computers, in Windows software for MS-DOS machines and, for example, in X-Windows for Unix-based machines. Because much of the information is presented on the screen, the user is prompted and reminded of the actions that are needed. Many of these GUIs employ some common conventions; there are usually menus headed 'File' and 'Edit' that access most of the basic file management and editing functions; these can be expected to operate in the same way in every program, and even on different hardware platforms. Another point worth remembering is that every menu item ending in an ellipsis (...) will lead to a dialogue box that, as well as indicating the range of operations that can be performed, will include a button marked 'Cancel'. Such dialogue boxes can be explored without fear of embarking on an unwanted operation.

A pull down menu and the dialogue box corresponding to the Open... menu item — note the ellipsis and the 'Cancel' button.

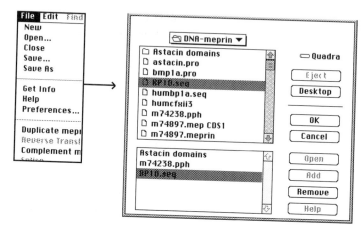

Software

Your choice of software will often be controlled by factors beyond your control, and you may be required to learn how to use several packages, running on different hardware 'platforms'. This may seem daunting at first, and there are many users who will never make much progress, reflecting the underlying attitudes that act as powerful deterrents to exploration. There are too many users who treat computer operations as a series of keystrokes (or mouse movements) that have to be written down and executed in exactly the same sequence each time. While this may be acceptable for tasks that are rarely needed, it is not an approach that will extend your skills beyond those specific operations. Indeed, you would not accept such a way of working for your laboratory research. It is far better to build a conceptual image of the intended task, and then implement that task without recourse to crib sheets and scripts. Sometimes you will choose a wrong option, and have to redo the task; however, if you analyse what went wrong you are unlikely to repeat the error. Try to conceptualize the process that you are going through, and interpret the computer operations in terms that you can understand: 'I want to save the file to take to another machine, so I will need the simple format (plain ASCII) saved onto the 3.5-inch disk (A:); but I must remember to save it in a rich format (program specific) so that I can keep all of the analysis details; I must remember to use a different filename for the two versions, because this stupid program doesn't warn me if I'm about to overwrite a file with another of the same name'. If at some stage you then start to use another program, you will retain the concepts, such that learning any new commands becomes secondary or trivial.

A second impediment to effective computer use is a reluctance to experiment and explore the software. You cannot break a computer by typing at the keyboard. The ellipsis tip is worth remembering; if you cannot recall the location of a particular menu option, pull down all the menu items until you find it. Experiment with non-critical data and at non-critical times. Scientific software can be much more fun to 'play' with than arcade-style games.

Perhaps it is wise to qualify this exhortation to experiment. Under the worst circumstances software will 'crash', which is a dramatic way of saying stop, or get hung up in an internal, never-ending loop. You will be unable to do anything with the mouse or keyboard, and you may have to restart the machine to proceed. These errors, at worse, usually lose your current data set (unless it was saved to disk as soon as it was entered — a lesson that you will learn, sooner or later, from personal and bitter experience!). However, there are two circumstances where the damage may be more widespread. The first is with software that can modify the storage of files on the disk; it is possible to erase a whole laboratory's data with a single command (although only a minor inconvenience if the data are backed up, this is not a good way to discover the extent of data back up in your laboratory!). Secondly, if your laboratory has several computers linked (networked) together, a crash on your machine may interfere with the work of your colleagues.

*Don't be tempted to use these books as a substitiute for the manual that will accompany a **legitimate** copy of the software!*

You may be daunted by the complexity of a new software package, to the extent that you never learn how to use it effectively. Many computer manuals are reference books rather than tutorials, and you often need to know how to use the program before you can read the manual! For 'business' software, this is reflected in the huge number of 'How to use… ' books — some of which are excellent. These types of books usually fill the shelves of the computing section of a large bookshop, and some may even be in your library. If you intend to use a package a lot, it is worth investing in one of these.

Scientific software

Scientific software packages are sometimes referred to as 'vertical applications', which is computer-speak for a program that performs a single task and which is used by relatively few people. Sometimes the software is available commercially, often it has been written by one or two scientists and is made available at no or nominal cost. For scientific software, there are two levels of program design that you must be aware of; the way in which you interact with the program, and the algorithms (or set of rules) that are used to achieve the analysis. While you may, as an end-user, worry only a little about the program design, you must be concerned about the algorithms that are used. If you use a program to fit a curve to a set of data, or to align two macromolecular sequences, it is essential that you understand the algorithm that is being applied, at least in outline. Search out and read the original papers that describe the algorithm or program. Read the manual! Explore the limitations of the software and, if possible, compare it with other programs. Devise sets of trial data with known answers and use these test data repeatedly, and especially if a new program is being used.

Most scientific software will offer the user comprehensive control over the analysis. This control is implemented by adjustment of program parameters, such as the window size in a sequence alignment, or the 'goodness of fit' in curve fitting. You will be expected to assume responsibility for these program parameters, even if you simply accept the default options that provide a reasonable starting position. Discover whether the program records or prints these parameters, and, if not, keep an independent record of them yourself. Many programs allow users to change the default settings; ensure that the previous user (who might have a different need, or be more or less knowledgeable) has not changed those defaults to values that are inappropriate for your needs.

A computer-based data analysis is one of the final, but nonetheless critical, stages of an experiment. It cannot be seen as a difficult, tedious or unimportant process to be left until you write your papers or thesis. Expect to invest a significant amount of time in learning how to use software, and how to use the analytical data that you derive. Learn to recognize when the analysis means that the experiment should be repeated, rather than wasting time analysing data that is fundamentally weak.

'Productivity' packages

In contrast to scientific software, most commonly used packages are designed to be general purpose and meet the needs of many users. With such flexibility, of course, comes a loss of specificity; it is unlikely that the package will do exactly what you wish, and you can expect to invest some time in learning how to configure the program for your own purposes. Indeed, in the early stages of getting to know such packages, 'productivity' may be rather low, as you spend more time finding out how to do something than you spend doing it. Moreover, many of these packages will have been designed to capture the 'business' market. You will recognize that the needs of business users and scientists are rather different, and will learn to discount inappropriate features. Sometimes, you may have to devise rather tortuous strategies to force the package to give you a suitable analysis or presentation.

Data-plotting packages

It is now rare to see researchers plot their data by hand, although the quick notebook sketch will always remain. Much of the production of graphical material — previously the domain of drawing pens, press-down key symbols and lettering stencils, has been replaced by computer programs that will, in the simplest instance, let you enter a set of data and present it as an annotated graph.

For numerical data presentation, look for a package that seems to have been written for science rather than for business. Business packages are easy to spot, offering many colours, histograms that can be assembled as piles of bananas, and pseudo-three-dimensional effects. They might fail to offer error bars, curve-fitting functions or the ability to transform data, such as 'molecular weight' to 'log (molecular weight)' for a DNA or protein size analysis. Scientific packages may look less exciting, but will generate graphics that are part of the universal scientific language. In particular, try to avoid those three-dimensional effects in histograms or line graphs; these look flashy, do nothing, get firmly in the way of data interpretation and decrease the 'data/ink' ratio.

Most packages offer default settings for data presentation; be wary of those options. Take particular care over axis scaling; if a data set runs from 42 to 100, it is not unusual for the package to elect to plot the data on a scale from 40 to 100 (or 40 to 120). This obscures the origin of the graph, which may be an important point that you are trying to make. You may want to superimpose, on a set of discrete experimental data, a curved line that represents the best fit of a particular function to the data (you will know the parameter values for the equation, such as

V_{max} and K_m for a Michaelis–Menten curve). If this cannot be done within the package, consider creating a closely-spaced theoretical data set in a spreadsheet (see below) to define the curve, import this data into the graphical package and plot it without symbols, but joined together.

Make sure that the package will allow you to save your data in a simple but widely accepted format — a tab-delimited ASCII text file would be the best option. Does the package support the best quality printer that you will use? Can the package generate pictures of the graphs in a format that allows them to be placed into another document (for thesis production, for example)?

Statistics packages

Most commercial statistics packages are extremely complex, offering analyses that range far beyond the needs of any one user. Responsibility for the use or abuse of these analyses rests with the user, and there is little virtue in applying a statistical test because it is available, when you know nothing about the purpose and design of the test. Thus, choice of a statistics package should follow on from a clear understanding of statistical principles and an awareness of the type of test that is apposite. For simple statistical tests, such as linear regression or descriptive statistics, you might even prefer to use the functions provided within a spreadsheet package.

Having specified your needs in terms of analyses, other factors can then be considered. Your main interaction with a statistics package will be through the data editor — is this comfortable to use but powerful? If the package is written for one of the standard GUIs it should be possible to 'cut and paste' data into and out of the program. If you already know how to use the GUI the learning curve will be much shorter than with a package that uses its own protocols and conventions to interact with the user. Does the package offer special features and analyses that you know you will require? If you will be conducting the same analysis time and time again, it is helpful to be able to store a series of operations in a 'batch file' or a 'macro'. Does the package support this facility? What support is available for the package? Perhaps you should use a package that is supported by your department or institution, as this may give access to a specialist who understands both the statistics and the software. The more people who are using a package, the more likely it is that bugs in the algorithms will have been eliminated — look for a package with a long history and with a reputation for good support by its suppliers.

Spreadsheets

A spreadsheet can be considered to be a two-dimensional array of 'cells' that can contain either data or 'formulae' that act on these data. For example, the cells from Row 1, Column B to Row 6, Column B could contain the numbers 2, 3, 4, 6, 8 and 10. Another cell, containing the formula 'SUM(B1:B6)' would display the value 33. If the content of B1 is changed from 2 to 10, the formula cell would be updated automatically to display the new number 41. This is a simple illustration, but the value of spreadsheets lies in the construction of 'templates', in which the relationships between data-containing cells are mapped out, and new data are entered and analysed automatically. From a simple calibration graph with unknown samples to a personal database of PCR primers that calculates melting temperature, G+C content and the molar extinction coefficient, spreadsheet templates have great value. If you intend to deal with a large amount of numerical data that are analysed in the same fashion each time, a spreadsheet template might be worth investigation.

Spreadsheets are also useful as 'quick and dirty' calculation tools, and can, for example, be used to generate closely spaced data sets for function plotting. Many of them have a complex programming capability (that generate macros, i.e. a series of commands that is applied automatically) and you might be able to automate routine analyses such that they can be executed with a single keystroke. The better spreadsheet packages offer sophisticated data-plotting functions, and a wide range of mathematical and statistical functions. Some include optimization algorithms that can be used to find the best fit

A sample spreadsheet. The content of cell B7 (which contains the formula SUM(B1:B6) is updated dynamically, as the contents of any of the cells B1 to B6 are changed. The formula in cell C7 is SUM(C1:C6).

File	Edit	Formula	Format	Data	Optio

B7		=SUM(B1:B6)

Worksl

	A	B	C
1		2	10
2		3	3
3		4	4
4		6	6
5		8	8
6		10	10
7	Total	33	41
8			

line to a set of data. All of these functions require that you spend some time getting to know the capabilities of the software; only you can decide whether this time is wisely invested.

Text editors, word processors and desktop publishing

It is unlikely that you will enjoy such luxuries as secretarial support, and you will be expected or may choose to produce your papers, reports and thesis using a word-processing package on a computer. Again, the choice of package is governed by local considerations; choose a package that is supported by a high degree of local expertise within the laboratory, department or institution. The high-quality packages will all offer complex facilities, but make sure, for example, that you know how to generate a table of contents or an automatic index well in advance of thesis production. You should not be experimenting with different packages or facilities in the middle of thesis writing; you should know that the package you are using will meet your needs before you begin.

Word-processing packages offer enormous flexibility in terms of the appearance of the text. A wide range of fonts and text styles and fancy graphical devices may be tempting, but you should aim for a subdued, consistent appearance to your written work when it is printed. Different textual characteristics should be used consistently to discriminate individual items in the text, such as body text, headings and figure legends. If your thesis uses Greek letters, try to find out how to incorporate and print them within the document; this saves the time and effort of adding them subsequent to printing. Laser or ink/bubble jet printers are high-resolution printing devices that can generate any image, such as a Greek character, given appropriate software.

The term desktop publishing refers to the integration of text and graphic elements into a finished publication. Undoubtedly, the production of a thesis qualifies for this description, and you may be able to produce a whole thesis by seamless integration of files containing graphics and text. The more sophisticated word-processing packages can import graphics quite well, although it will be helpful to be able to see the finished product on screen without printing. However, programs written specifically for this task are more powerful, and usually emphasize the importance of seeing what is being created. If you intend to use such software, give early consideration to the formats in which you will store your data. Text and graphical files written to disk by one program may not be readable by another. Take particular care if data are to be derived from instrument control software. These often produce, by default, files in internal formats that, though efficient, are non-standard and unlikely to be readable by other programs. You may need to save the data files in a simpler format, such as a tab-delimited ASCII file.

Data-presentation packages

Data-presentation programs focus on the visual representation of numerical data, but other packages might generate high-quality pictorial images of DNA or protein sequence data, or provide general purpose drawing tools. In an integrated software environment such as that provided by a GUI, the images generated by these packages should be inter-convertible and transferable from one program to another for embellishment; such modifications can overcome the deficiencies in most packages. Remember though that the process of scientific graphic design should precede, and is external to, production.

Many scientific packages can produce text-based analyses, or graphical representations thereof, and you should consider the nature of the data that you want to retain. Programs that run under one of the GUIs are relatively seamless in their integration, and a graphic generated in one program can usually be 'cut' and 'pasted' into a second, such as a desktop publishing package. The graphical image will demand a lot of disk space, and it may be preferable to store the raw data and recreate the graphic when it is needed.

See Chapter 3, 'Conducting a research project'

Alternatively, the graphic can be generated, printed and stored away for eventual use in a paper, report or thesis. This would mean that you would have to be consistent in the graphical output that you use from the earliest to the last analyses. You would in any case only use a fraction of your data in your thesis, and while the graphic image will be valuable in interpreting the experimental data, it is preferable to store the raw data and create the final images when needed. The advances in printer technology and in printer-driver software mean that you will be able to produce your graphics using the best facilities that are available at any particular time. If you have ever tried to read a DNA sequence that was printed with a dot-matrix printer, and compared it with laser-printed output, you will appreciate the benefits of developments in printer technology.

Electronic mail and networks

Most mainframe or large, multi-user computer systems are connected to national networks, which in turn are connected to international networks. Scientists can, from the computers in their offices, connect to the computers of colleagues all over the world. What benefits ensue from becoming part of this scientific network?

Computer afficionados sometimes refer to 'snail-mail' — an archaic system of communication that requires envelopes and postage stamps!

— you can contact another computer user anywhere in the world

— you can subscribe to scientific bulletin boards and newsgroups

— you can access at least 100 databases of scientific data

— you can obtain public domain scientific software.

To avail yourself of these facilities, you have to learn how your computer is connected to the rest of the world, and how to make contact with this community of users. The computer will be a large, multi-user system, although you may interact with it via a personal computer functioning as a terminal. Usually, you will need to apply for authorization to use this multi-user system; registration will normally give you access to courses and documentation.

In addition to basic activities such as file management, other processes that you will need to understand are electronic mail, file transfer and, possibly, newsreading procedures.

When you register as a user of a mainframe computer, you will normally be given a unique username that identifies you exclusively. If you are lucky, your username will be something memorable such as 'beynonrj' but your computer system may equally well decide to christen you 'sbty567'. In combination with a password, which only you should know, you will use your username to 'log on' or connect to, the mainframe.

All of the e-mail addresses in this book have been changed to non-existent computers or users — it is not even worth the effort of trying to connect to them!

In turn, the computer will itself have a name that identifies it uniquely within the country and sometimes also within your institution. A machine name such as 'liverpool.uxq' identifies a specific computer (UNIX system Q) and the city (Liverpool). Finally, two additional prefixes ('uk.ac') added to the address indicate the country (U.K.) and nature (Academic Community) of the site. A full international address would therefore be beynonrj@uk.ac.liverpool.uxq. Another user, anywhere in the world, should be able to send e-mail to this address. In practice, life is not this simple. There are several address conventions and it can be difficult to work out the correct address form. Such difficulties can make your first exposure to e-mail rather frustrating. If you feel that the benefits of e-mail outweigh the difficulties of learning, the following strategy should help you to get started.

- Attend a course or acquire the user manuals for the mail system on your local computer. This will introduce you to the commands and initial skills that you will need. These include the ability to use the text editor to enter messages, to read, file and reply to an incoming message, forward a message to another user and send a message to more than one user at a time.

- Understand that the computer systems distinguish absolutely between SBO6 and SBØ6, although both are usually spoken in the same way; 'S-B-oh-six'. It is helpful if the computer system generates the number zero with a slash through it, such as in this example. Similar confusion can arise with the number '1' and the letter 'l'.

- All mail systems will have a 'reply' command that will let you respond to incoming messages. The mail system will work out the correct return address from the incoming message. Thus, if you cannot work out a valid address for a person, ask them to send you a message. When you 'reply', make a note of the address that is being used.

- Learn a little about address conventions and rules. This knowledge can often help you to work out a valid address. Cite your e-mail address in regular paper-based correspondence. Such practices will encourage others to use this form of communication.

Sending data by computer   Facsimile (fax) machines offer some of the advantages of e-mail, being relatively cheap and immediate, but as a communication device, fax machines have disadvantages. It is quite likely that somewhere in the world, a 2-kilobase DNA sequence has been read from sequencing gels into a computer, printed, and faxed to another scientist, who then re-typed the sequence into their own computer. Tedious, time-consuming and error-prone! The same sequence could have been embedded in an e-mail message and sent error-free to the recipient. As well as sequences, it is possible to use e-mail to send images (a gel, for example) complete manuscripts and almost any type of computer-encodable information. It is usually necessary to convert complex files into an encoded format that is compatible with e-mail systems. This encoded file can then be transferred to the mainframe, embedded within an e–mail message and sent to the recipient, who extracts the embedded, encoded file, then decodes it and recovers the original file. This works well, and there are many public domain programs for encoding and decoding binary files. Seek advice from your local computer advisory service or from local 'gurus'.

Mail servers

Electronic mail is usually sent to another human, but scattered around the globe are 'mail servers'. These are effectively computer programs to which you send mail. They respond automatically, usually by sending you information. For example, the European Molecular Biology Laboratory operates a mail server that includes sequence data and molecular biology software. The address, name@de.embl-heidelberg is not a person, so you cannot send a free-form text message ('Dear people at EMBL, please send me a copy of DottyPlot'). The grammar and syntax used by such servers is very restricted, and you will have to learn how to formulate your requests. In such cases, it is a good idea to start by sending a message containing one word, 'help' (no quotation marks) to the server. These servers are a valuable resource, and you should be aware of their contents at least. You could use them to obtain a machine-readable form of a sequence that has a Genbank accession number quoted in a paper, or to retrieve a three-dimensional co-ordinate file for display on your local facilities.

Other mail servers provide access to DNA and protein sequence databases. If these servers are sent an appropriately formatted message — one that includes a DNA or protein sequence — they will search the database and reply with a mail message that lists the sequences within the database that bear the greatest similarity to the sequence that was sent in the original message. The main advantage of these servers is that they can be used by anyone, need no local software other than e-mail, and they search databases that will have been updated the previous night (local databases, such as those provided on CD-ROM, can lag several months behind the current versions).

Bulletin boards

Another type of e-mail system is the bulletin board (also known as a 'bboard' or 'list'). Bulletin boards are public, and you, or anyone else, can 'post' a message there. In turn, the bulletin board is read by a large number of subscribers. They can 'followup' your message, in which case the reply is read by all members of the group, or they can 'reply' directly to you, in which case your dialogue is private.

:-)

;-)

:-(

Append one of these triplets of characters to e-mail messages where appropriate!

In the life sciences there are a number of bulletin boards of interest (see the bionet list adjacent — the number alongside the name indicates the number of articles at the time the list was prepared). There are two ways to gain access to these bulletin boards. In the first, you subscribe to the bboard or list and all messages are sent to you directly, via e-mail. This is not the recommended method, because you will receive tens of messages a day. Take a few weeks holiday, and your e-mail software may give up at the prospect of handling the backlog! Also, if ten users at one site all subscribe to the same bulletin boards, the same messages are sent to each user — a waste of 'bandwidth' that stresses the (finite) capacity to move electronic data around the world.

It is quite likely that your institution's computer system has the capability to receive and process 'news'. This software will receive one copy of all news items and make them available to all local users without replicating further copies. Be warned; there are over 400 newsgroups, and only about 30 of them are strictly biologically based (such as those listed alongside) and it is easy to lose time browsing through groups such as 'alt.beer'. Learn how to tell the news software that you are only interested in a small subset of the newsgroups.

The newsreader software should allow you to post, 'followup' or reply. When posting or following up, your message will be read by many hundreds of subscribers, and there is an obligation upon you to observe certain rules, or 'netiquette'. In the absence of verbal cues, statements meant to be tongue-in-cheek can be misinterpreted and cause irritation. A mail message that is deliberately offensive will guarantee an electronic 'flame'; a flurry of messages, some public, some private, that make clear that you have overstepped the rules of the electronic community. If you want to indicate the tone of a comment, append the most appropriate of the triplets of characters alongside (look at them sideways to see what they mean).

The diagram overleaf is a 'screendump' (a captured image of what was on the screen) of a 'news' session where the 'Methods and Reagents' bulletin board was being perused. The list of topics indicates the range of subjects, and the list of names/e-mail addresses indicates the originators of the messages — although not clear from this screendump, they come from all over the world.

```
             bionet.molbio.methds-reagnts (68U 0K 0H)        You have mail

      61        Immunoprecipitation                     Howard Kirk
      62   +    pXM vector sequence                     droopy@vthvax.tamu.e
      63   +    electroblotting southerns               Philip Swallow
      64   +    MacVector                               sleepy@yang.earlha
      65        screening cDNA libraries                grumpy@ucbeh.san.uc
      66        Sequencing survey                       Morris Zapp
      67   +    suscription                             Robin Penrose
      68        RNA ligase & single-stranded DNA        Eric C. bashful
      69   +    yeast cDNA library                      sleepy@UTBC01.CM.
      70   +    cDNA libraries (Reference)              bashful@ucbeh.san.uc
      71   +    promoter sequence listings              Henry Beamish
      72   +    Lab stocks database: re-re-re-inventing the w   Henry Beamish
      73   +    help                                    s0679400
      74   +    help                                    meprin@liv.
      75   +    pET9 vector sequence                    dopey@mcclb0.me
```

In turn, any one of these messages can be viewed in full, as with the example below. Having read this article, you can learn from it, particularly if it is a reply rather than an original posting, reply to it, 'followup' yourself or ignore it. Indeed, some of the bulletin boards are so active that it can be difficult to isolate articles of interest from the background 'noise'. Good newsreading software will allow you to search for specific words in the subject line.

```
14 Aug 92 17:04:15 GMT    bionet.molbio.methds-reagnts    Thread  68 of  75
Article 117            RNA ligase & single-stranded DNA      No responses
xxxxxx@bioscience.xxxx.edu  Xxxx X. Xxxxxx at Dept. Biology, University of XXXX

I am going ligate oligonucleotides into linearized single-stranded plasmid
DNA using RNA ligase. When I last did this (years ago), it was very
inefficient. I would appreciate any favorite enhancements or references
to this method.
_____
|  Xxxx Xxxxxx                          Dept.Biology, U. of Xxxx        |
|  "Being certain of the conclusion assists in finding the proof"       |
|                                       Galileo, 17th century           |
|_____|
```

File transfer

As mentioned previously, e-mail can be used to send encoded files to another user, but this is not particularly convenient. In particular, the size limitation on e-mail messages means that encoded files sometimes have to be split, sent in parts and assembled with a text editor before decoding. An alternative is to send the file to the recipient as an unencoded binary file. To do this, you will have to become familiar with file transfer protocols, or 'FTP'. FTP can be complex, and you should consult your local computer service for advice. One type of FTP that is somewhat simpler is 'anonymous FTP' in which you would connect to a remote computer, which may be on the other side of the world, browse through public directories, identify files of interest and retrieve them to your local filespace. These files can be of many types: pictures, software, text files, sequences and so forth. You are allowed access to a restricted part of the remote computer, but cannot change or delete files. In effect, all you can do is browse and retrieve. If you have anonymous FTP you can retrieve files that are in one piece and not encoded (binary), although they may be compressed to reduce the disk space that they occupy. You will need to use software to decompress or 'unpack' the file before it can be used.

Gophers

Gophers ('go for') are an exciting recent development that allows access to a wealth of data from your terminal. They are easy to use, and provide sophisticated facilities, including the ability to download programs, pictures, DNA and protein sequences and protein structure coordinate files. They are efficient, because they only connect to the remote computer (the Gopher 'server') transiently while you ask for a list or a file. There are many specialist Gophers, and it is easy to become lost in 'gopherspace'. To use Gopher, you need a relatively sophisticated and fast connection (Ethernet) to the rest of the world, but is worthwhile asking your computer gurus if they can provide you with Gopher 'client' software.

Caveat

A final word of warning. It is difficult to find the right balance in the use of computing in your science, and there are some students who become so enthused by computers and computing that they no longer give adequate time or attention to experimental work. If you find that you are spending time in front of the computer instead of doing laboratory work (and you will know, in your own mind, when that is the case) it might be a good time to review your position.

Further reading

Bishop, M.J. & Rawlings, C.R. (1987) 'Nucleic Acid and Protein Sequence Analysis — A Practical Approach', IRL Press, Oxford (Eds.)

Bryant, T.N. & Wimpenny, J. (1989) 'Computers in Microbiology — A Practical Approach', IRL Press, Oxford (Eds.)

Bryce, C. F. A. (1992) 'Microcomputers in Biochemistry — A Practical Approach', IRL Press, Oxford (Ed.)

Fraser, P.J. (1988) 'Microcomputers in Physiology — A Practical Approach', IRL Press, Oxford (Ed.)

Ireland, C. R. & Long, S.P. (1984) 'Microcomputers in Biology — A Practical Approach', IRL Press, Oxford (Eds.)

There are many books that introduce computing and computers. The Practical Approach Series includes these five books on computer applications in the life sciences. Although the individual chapters vary in the level of treatment of their respective subjects, these books are a valuable introduction to much of the jargon, and to a great deal of further reading on most of the topics alluded to in this chapter.

Safety matters

Developing a safety awareness

A research laboratory is a dangerous environment, and the only reason why accidents are so few and far between is because those who work in such an environment are aware of, and react appropriately to, the hazards. It is likely that you will have spent some time in a research laboratory as part of your undergraduate training, but you may have been shielded from many of the risks. All this changes when you embark upon a research programme — you will be expected to take much more responsibility for the day-to-day running of your research, and to build an awareness of hazards and the correct actions to take. Do not take this chapter as your source of safety information — make sure that at the earliest opportunity you embark on the fact-finding and training that you will need.

In a well-run department and laboratory you will enter an environment that is acutely safety conscious, and which will have a well-established structure to provide you with advice, training and the appropriate regulatory approvals. Your arrival should automatically initiate a series of events that ensure that you are as well protected as is possible. This may take the form of safety seminars, of interviews with trained individuals and of discussions with your supervisors and others in the laboratory.

Much of safety awareness is common sense. If you take the time to pause and look around any laboratory, you will see situations that are potentially hazardous. Simple things like poor lighting, a leaking sink that puddles water on the floor or a frayed power lead seem to be so obviously hazardous that no one notices anymore! Also, it is surprising how often such common sense is abandoned in the 'heat' of an experiment! The best way to ensure that procedures are carried out safely is to prepare for as many contingencies as possible, which means applying common sense in advance.

You will be best protected, as will others around you, if you can retain this common sense, combine it with a precise knowledge of the nature of the hazards that confront you and have a clear-headed response when things go wrong, as they inevitably will. This knowledge comes from a critical review of the procedure and a confidence gained from a 'dry-run' where appropriate.

A third component of safety awareness is good housekeeping. An uncluttered environment is important, and the more hazardous a

Three components of good safety awareness:

- *Common sense*
- *Good knowledge*
- *Good housekeeping*

procedure, the cleaner and tidier the environment should be. For example, in the event of a spillage you cannot easily decontaminate a fume hood that is full of bottles and equipment; floors should be free of equipment and boxes that might cause you to trip and fall when carrying a dangerous chemical, or glassware. You footwear should not slip on the laboratory floor and should protect you from spills. Aim to be mentally and physically comfortable in your working environment.

Whose responsibility?

It is easy to assume that others will take all of the responsibility for your training, that they will be responsible for your protection and that they will ensure that you have satisfied any legislative requirements. Ultimately though, it is you who will use the radionuclides, run the electrophoretic separation or inject the animals. You therefore have a responsibility (that extends to a legal responsibility in many circumstances), and a right to appropriate training, awareness, advice and permissions. But, it is easy for your research to move in directions that were not anticipated, and for your work to move into new areas for which new approvals or training are needed. When your work takes a significant change in direction, establish that new hazards are recognized and properly controlled.

Many laboratory procedures will be given to you as *faits accomplis*, but this is no intrinsic reason to assume that they are safe, to you or to others. Moreover, if you adopt safety precautions without considering why they are in operation you have abandoned the common sense awareness that is so important. If you adopt 'standard' procedures, take time to review them, and convince yourself that they are safe.

It will not normally be your responsibility to correct any violations of good laboratory safety, other than for your immediate responsibility to make the area safe for others. You will be expected to report such violations, however. Do not assume that 'someone else' has reported the faulty mains lead, a patch of water on a slippery floor, a failed light bulb in a dark stair well. It is likely that 'someone else' assumed that you or another 'someone else' reported the hazard. First, make the area secure, and then pick up a telephone and call the person who is designated as being responsible for correction of such a hazard — it should be easy for you to find out who this person is. You might prefer to leave the person responsible a note, giving the time and date, location and nature of hazard, and your name and telephone number. Just as you have a responsibility to report faults, so others have a responsibility to act on your report. Do not let this matter rest until the hazard has been removed.

Occupational health problems

If you are unlucky, you may encounter some of the many work-related health problems that are a feature of virtually all working environments e.g. back problems caused by lifting heavy equipment or from poor seating at a computer, a hypersensitivity to protective surgical gloves or an allergy to experimental animals could all beset you. It is essential that you report such problems, consult the institution occupational health physician and find solutions — do not 'grin and bear it'.

A change in your body can put you at risk from new hazards. An illness, or a long-term drug treatment can change your susceptibility to chemical hazards. Pregnancy brings with it new risks from teratogenic chemicals and ionizing radiation. Maybe your eyesight is not so good, and you can no longer read a pressure gauge easily? Often you will be the only one who knows about such changes, and are therefore the only person who can ensure that you remain safe. Do not delay in making enquiries — no additional experiments can be worth more than the health you are compromising by performing them. Tell someone immediately. You can expect complete confidentiality and sympathetic treatment, and solutions will always be found.

Peer group pressure

In your laboratory there could be other scientists who appear to have abandoned their safety awareness. There can be no justification for any relaxation of safety procedures, and you should resist any peer-group pressure to follow suit. You will come across many of your peers who eschew the protection of laboratory coats, who pipette by mouth, or who '*never use gloves handling tritium — it's not dangerous enough*'. These individuals are foolish, often contravene legislation and should not be adopted as role models. Do you really want to accept part of a chocolate bar that is sitting on the same bench as a bottle of a carcinogen? Extreme examples perhaps, but it is safe to bet that sometime, somewhere, this has happened. When it comes to safety procedures, you must know your own mind, and have the courage of your convictions. If you cannot take responsibility for your own safety, what guarantee is there that you will consider the safety of others?

General laboratory safety

One of the most commonly occurring laboratory accidents are cuts caused by glass. Indeed it seems obvious that an item of broken glassware will have a sharp edge that should be immediately discarded or repaired (even if the cut surfaces are simply flamed smooth). Yet, a quick perusal of many laboratories will reveal glassware in a state of disrepair that is dangerous, and about which nothing has been done. Dispose of broken glassware in specified containers, according to local practice — many cleaners have been cut by broken glass thrown into a waste bin not intended for that purpose. The same argument applies to disposal of 'sharps' — needles and scalpels — especially when these have been in contact with biological samples.

Equipment powered by mains electricity abounds in the laboratory, as do large quantities of conducting solutions (salt solutions, buffers, etc.). Indeed, during electrophoresis, a DC voltage and an electrolyte are brought into close proximity, needing only the addition of a careless individual to complete a lethal circuit! The safety interlock devices are designed to preclude this combination, and should not be defeated, no matter how inconvenient they are.

Frayed mains leads, faulty plugs, connectors that get warm, arc or smell of burning must not be used. It is inconvenient waiting for a technician to investigate and repair the fault, but not as inconvenient as completing your research project in a prefabricated temporary laboratory while your building is being rebuilt! Do not be tempted to effect repairs yourself. Discover the locations of fire extinguishers in your laboratory, and ascertain the types of fires against which they are effective.

There will be individuals in your department who are qualified to administer first-aid. Find out who these individuals are, and where they work — they can prevent a minor incident from becoming a major one. Indeed, consider attending a first-aid course yourself — you may save a life one day. Know the whereabouts of first aid stations in your laboratory, and familiarize yourself with the contents.

Chemical hazards

Chemical hazards will vary from laboratory to laboratory, and from procedure to procedure. No general guidelines can be given over and above those that fall into the common sense/knowledge category. In the U.K., the Control of Substances Hazardous to Health (COSHH)

regulations require that you assess the hazards inherent in the use of a chemical before you use it. In many circumstances, the amounts of a chemical that will be used are small, and no specific precautions will be needed. Under other circumstances, the potential hazard may be such that very stringent procedures and operator protection are called for. Most important is that you are aware of the appropriate response, which means that you must be in a position to make an intelligent assessment. Thus, *all* chemicals used in *any* experiment must be assessed as needing standard precautions or according to the demands of a specific assessment. The factors that you should consider in this assessment are:

— toxicity/corrosiveness

— instability or inflammability

— quantities to be used

— volatility

— dustiness of the chemical/ease of inhalation.

Be particularly cautious if you scale up an experiment; the new quantities of materials can introduce additional hazards.

Other aspects of chemical safety are sometimes overlooked. If a hazardous chemical is transferred from one container to another (perhaps you will acquire some of this material from another laboratory) the hazards and hazard pictograms should be attached to the new container. Weighing balances are notorious for the snowdrift of anonymous chemicals that surround them — it is your responsibility to clear up spillages immediately as you are the only one who knows the identity of the chemical and the correct clean-up procedures. Dispose of unused materials correctly, according to recommended procedures. This is particularly true when you leave a laboratory, if the materials are no longer needed.

Radiochemical hazards

If you use radioactive materials in your research, you will need to develop a good awareness of the types of hazard that you might face, and be trained in the relevant procedures. Your department will have a person nominated as the radiation protection officer who should be

Isotope	Half-life	Emission	Shielding
^{125}I	59.6 days	35 keV γ	>2 mm lead
^{51}Cr	27.7 days	0.32 MeV γ	>2 cm lead
^{3}H	12.4 years	18.6 keV β	none needed
^{32}P	14.3 days	1.7 MeV β	1 cm Perspex
^{33}P	25.4 days	0.25 MeV β	none needed
^{35}S	87.4 days	0.167 MeV β	1 cm Perspex
^{14}C	5730 years	0.156 MeV β	1 cm Perspex
^{45}Ca	163 days	0.257 MeV β	1 cm Perspex

The radionuclides used commonly in biological research.

consulted before you work with any new isotope or radically different procedure. You will need to register as a user of radioisotopes, and may need to give a blood sample. Some of the more common radioisotopes used in a biological laboratory are listed above. The biological hazard associated with each isotope is governed by the chemical nature of the radiolabelled compound, the nature and energy of emitted radiation and the tendency of the isotope to accumulate in specific tissues or organs. Note that one of the most commonly used isotopes, ^{32}P, is also the most hazardous.

The most immediate danger when handling radioisotopes is that of radioactive contamination of yourself, of others or of the workplace. In all instances, individuals are now exposed to a radioactive material without being aware of the fact, will be unprotected and may ingest or distribute the radioactive material further afield. As with any hazardous procedure, the chances of errors are minimized if you are thoroughly familiar with the principles of what you are attempting and the procedure itself. Anticipate the sorts of problems that you might encounter — halfway through a radiolabelling procedure is not the time to discover that you need a new rack of pipette tips! The level of protection needed for most radionuclides encountered in the biological laboratory requires devices such as shields, gloves, and fume hoods to extract volatile materials. These procedures are not so restrictive that it becomes inconvenient to conduct dummy runs of all procedures, replacing the radioactive material with non-hazardous materials. Produce a detailed checklist for the materials and steps of the procedure and use it.

Most radioactive molecules are packaged in such a way that accidental damage is minimized, and any material that escapes is retained. Some of the packaging methods are ingenious, but can be tricky to manipulate. It is worthwhile investigating a container that no longer contains large amounts of radioactive material. Although it will still be radioactive, it will allow you to examine the packaging with less risk.

High energy emitters can be detected by badge dosimeters, but these are normally collected and screened at long intervals, and are therefore unsuitable for immediate assessment of contamination. You will not know whether you are in a contaminated environment unless you perform frequent monitoring, whether with a hand-held monitor or by swabs that are subsequently counted in a scintillation counter. If you encounter contamination, secure the area, and clean up according to standard procedures until monitoring reveals that the area is safe. If you become contaminated yourself, it is essential that you seek immediate help and advice. All contamination incidents must be reported to the radiation protection officer.

Disposal of waste radioactive materials is governed by regulations. There will be detailed local procedures for waste disposal — you must adhere to these rigidly.

Biological hazards

Your work may introduce the potential for biological hazards. You may be handling organisms that are pathogenic, or tissue samples (especially human) that have the potential to contain pathogenic material. Your department will have strictly regulated procedures for such work, and you must adhere to them. Special areas of the department will have been designated as suitable for this work, and will have safety equipment that provides the appropriate level of protection. Specific codes of practice exist for handling micro-organisms, human tissues or oncogenes, and you must be aware of these codes, understand their purpose and adhere to them.

If your work involves specific pathogens, or human tissues (including blood) it may be necessary for you to be immunized against specific pathogens, such as hepatitis B. You should be appraised of such requirements before you start work.

A second type of biological hazard might seem rather unreal, but the ease with which recombinant DNA work can be conducted introduces the potential for construction of new organisms with altered biological properties that may be detrimental. In the U.K., the Genetic Manipulation Advisory Group has developed an assessment procedure that is based on three factors:

- Access — the probability that the manipulated organisms can enter the human body and survive

- Expression — the anticipated level of expression of the inserted DNA

- Damage factor — a measure of the risk of a gene product having detrimental effects.

However fanciful you judge such worries, you must be prepared to make a detailed assessment of the likely risks, and to adopt the safety procedures and containment facilities that are appropriate for the work you are conducting.

You should not really work on materials, and never on live cells, taken from yourself. If your cells became transformed in a laboratory procedure (potentially to tumour-causing cells) you would have no intrinsic immunity to them.

Disposal of biological materials must follow recommended sterilization procedures, as must treatment of laboratory equipment and laboratory coats.

Experimental animals

If your research requires that you use experimental animals, your work usually falls under a well-defined and restrictive legislation, the objective of which is to maximize the well being of the animals. Indeed, you will find that there are virtually no procedures that you can conduct on an experimental animal without a licence issued by a regulatory, usually governmental, body. Your supervisor will advise you if you will need such a licence, but because the preparation of an application and award of a licence can take some time, it is preferable to make an early application. Until the time when your licence is awarded, you will be dependent on other licenced individuals in your laboratory to conduct any procedures that you need.

Coda

It couldn't happen to me

Safety awareness is often dismissed as being 'boring' and irrelevant. Most accidents happen when individuals are ill-prepared. Rather than taking time to discover the hazards that they face and how to deal with them, such individuals assume that they are immune to the danger. Certainly, serious laboratory accidents are rare, but it is equally true that the reward for the laboratory work of some scientists has been a fatal accident or a terminal illness. These individuals didn't think it could happen to them, either.

Author's note

I am uneasy about this Chapter. Unlike the rest of this book, which offers guidelines but allows you to decide what to do, laboratory safety is not optional. You must adhere to the procedures and regulatory requirements that apply in your department, institution and country. I am not qualified to give safety advice, and all I can do here is to indicate the sort of awareness you should aim to develop. You must seek specific advice within your department. Most importantly, recognize that the responsibility for your safety rests first and foremost with you, even if legal responsibility lies elsewhere.

No reading list is given here either. Your department will provide you with all of the information and material that you will need — all I ask is that you read it and act on it!

The graduate student as teacher

Introduction

Your department may be responsible for teaching hundreds of under-graduates every year. In the biological sciences, an essential component of that education is a training in the practical skills that underpin the subject. Lectures and tutorials complete the formally structured teaching. As a graduate student, you may be invited to participate in this teaching, either by demonstrating in practical classes or sometimes through tutorials. Remember your own experiences in tutorials and practical classes, and recall the postgraduate demonstrators who were responsible for teaching you. You are now one of those teachers.

Why should you invest the time and effort to become a good post-graduate instructor? You may welcome the opportunity to experience teaching before you decide on an academic career. It is a good opportunity to develop your abilities as a teacher and to talk to a group of students about your subject. Moreover, to teach a subject you will need to know it well; the level of understanding that you acquire will stand you in good stead in the future. Usually, there is some remuneration, which can go some way to relieve the stress of living on a postgraduate stipend.

Demonstrating in practical classes

The two extreme views of demonstrating — a boring, time-consuming chore or a valuable opportunity to teach and learn — are largely dictated by your attitude. The demonstrator (remember your under-graduate days?) is commonly seen in a passive role, sitting at a front bench and only 'speaking when spoken to'. However, your contribution can be much greater than this.

Organization

It is difficult to generalize, but you will probably find yourself either in a class with specific responsibility for an experiment, in which case the students undertaking 'your' experiment will change every week, or you will be given charge of a group of students throughout the course, for a whole series of experiments. This choice is usually dictated by the personal preference of the course organizer. In the first instance, you will get to know one practical very well and will be able to trouble-shoot that practical with a high degree of competence. In the second situation, you will get to know one group of students and be more able to assess their progress. Keeping track of the written work tends to be simpler.

Before you begin to demonstrate in any class, you should have a preliminary discussion with the course organizer and, if possible, sit in on any introductory lecture to the undergraduates. If your course organizer does not have such a preliminary session, consider asking for one, as it can be helpful to clarify individual difficulties with practicals and assessment procedures. If you have responsibility for one experiment, perhaps you should run through the whole procedure before the course begins, following the instructions as an undergraduate. This can highlight potential problems, and give you a much better feel for the experiment.

You should be in the teaching laboratory sometime before the class is due to start. This means that you can start promptly, become aware of potential problems and prevent students starting before their peers and before your introduction. It is helpful to your charges if you wear a name badge at all times in the teaching laboratory. The undergraduates will usually be given a name badge that they too must wear. Just as you would have appreciated a demonstrator who knew your name, try to get to know your student group as well.

Give your group an introductory talk on the practical that they will be doing; emphasize potential hazards and problem areas and ensure that they have read the schedule in advance. Naturally, you must have read and understood the schedule as well! Throughout the class, try to walk among your group and keep an eye on their technical abilities and procedures. Correct them before they make a mistake and periodically quiz them about the purpose of a particular step or experiment. Make sure that the undergraduates know what they are doing and why.

If you are ill or have to be away from the department for any reason, it is helpful, and courteous, to let the course organizer know as soon as possible. Could you swap with one of your colleagues for that one week? If you do so, make sure that they are briefed on the experiment. It is probably better if you continue to assess the write-ups, for consistency.

Although you and the students should be given explicit instructions about the nature of practical write-ups, you will still come across a wide range of efforts, from the many-paged discourse that repeats all of the practical manual, and which is far too long, to the scruffy, one-page effort. Could the latter ever be worth more marks than the former?

Undergraduates are given very little training or help in the generation of a practical report. As the person who sees their efforts, you must guide them, and make sure that they can record and analyse data, and express their conclusions, as well as they are able. Correcting each report in detail would be too time consuming, but why not take the group to one side and outline your ideas about a good write-up?

See Chapter 7, 'Safety matters'

Safety   You will assume some responsibility for the safety of the undergraduates in the teaching laboratory. You must be seen to be obeying the letter of the law at all times:

- Wear safety spectacles where appropriate, unless you wear spectacles normally; even then, you **must** wear them for dangerous procedures.

- Wear a laboratory coat at all times, properly fastened. No undergraduate should be allowed to work in a teaching laboratory without a laboratory coat. These laboratory coats should not be worn outside the class, and never in communal refreshment areas.

- Make sure you know the whereabouts of emergency exits, first-aid kits, wash bottles, emergency showers and fire extinguishers.

- You must also ensure that the undergraduates have read and obey the safety regulations. They may even be asked to sign the instructions to certify that they have read them.

- It almost goes without saying that smoking is not allowed. In addition, eating, drinking and even chewing gum in the laboratory must not be permitted.

- You must discourage all forms of horse-play in the teaching laboratory. Try to keep the aisles between benches clear of bags, coats and other extraneous junk; it is easy to trip over such items and they impede rapid escape from the laboratory.

- Act firmly and do not feel that it is a betrayal to report an errant student to the member of staff in charge; one small accident prevented is sufficient justification.

Feedback At the end of a practical course you will have a pretty good idea about the strengths and weakness of that course — or of the specific parts of it that you have taught. There should be a '*post-mortem*' session where all teaching participants have an opportunity to comment on various aspects of the course. The undergraduates will have identified with you more than the academic staff, and communicated to you both their positive and negative opinions of the course. Additionally, your own opinions will be valued. Make notes as you go through the course. Do not be afraid to suggest changes to organization and procedures, to the manual, or to any other aspect of the course. Remember, if you are unsure about something, or even uncertain about a particular matter, take it up with the course organizer.

Tutorials The tutorial should have been designed with a clear objective which, ideally, will have been shared with you and the undergraduates. During the tutorial, most of the talking should be done by the undergraduates; a tutorial is not a lecture. To conduct a successful tutorial, you are likely to work much harder than you do in demonstrating. The students cannot embark upon displacement activities such as labelling test tubes, and they may be shy, reticent or even antagonistic to the idea of expressing their ideas in front of their peers. It would be unfortunate if you were asked to conduct tutorials without some form of training, but that remains a possibility. If so, you might want to consult one of the many books on small-group teaching and develop your own ideas — you and the students will benefit from your efforts. Some of the key problems and possible solutions are highlighted below.

■ Your students will not know, and may be reluctant to address each other. Start the first tutorial with each of you introducing yourselves. Encourage the students to discuss points among themselves, rather than with you.

■ Nearly all small-group activities require the students to prepare material in advance. If none of the students has prepared their material, you may be the only one who says anything. If this happens, remind your group about the work that they will have to prepare for the next tutorial, and suggest that you will pick somebody at random to go through their work.

- Your tutees may ask you something that you do not know. Do not bluster or pretend that you do. You will be respected more if you admit your lack of knowledge, and use this as a means of planning, with your tutees, how you will find the answer. Have a brainstorming session to cover all possible information sources, and then send your group off to find the answer before the next tutorial. Of course, in the meantime you must also discover the answer. Get your group to discuss the effectiveness of the different ways in which they approached this problem. You will have turned a potentially embarrassing situation into a strong learning experience.

- Your tutorials may be rather formal events, with little evidence of the development of spontaneous discussions. You might try embarking on 'rounds', where each student **must** say something in turn. For example, they might be asked to complete a sentence such as 'Properties of restriction enzymes include…'. If you use this device, you must allow 'pass' as an answer.

- If you really feel that your tutorials are not going well, why not have a round on 'Why we're not working well as a group'.

Assessment

At the end of the course, the practical or tutorial marks are collated and used as part of a final assessment of the undergraduates. Implicit in this process are two assumptions: that all assessors mark identically and that we all have similar views on the merit of a particular mark. It is imperative, therefore, that course organizers tell demonstrators and tutors what scheme they are using and how to interpret the mark categories. A practical course can account for as much as 30% of the course assessment, and it is essential that such a large segment of marks be apportioned fairly and accurately.

"My impression is simply that you're not using our elegant marking scale, with its plusses and minusses, and query plus minusses, with quite the delicacy you might."

Malcolm Bradbury,
'The History Man'

Unfortunately, there is no uniform or satisfactory marking scheme for practical classes. Individual course organizers will probably have evolved their own specific marking schemes that may vary from a scale of 1 to 10, a scale of letters from A to E, a scale of letters including 'pluses' and 'minuses', a five point 'excellent, good, fair, poor, terrible' scale or a gold star system, etc. This is basically a mess and is not helpful to a demonstrator who may teach in two different classes, using different schemes. Is a 'C' equivalent to 5, 6 or 7? What is the difference between a B-plus and an A-minus?

If the department thinks that your assessment of undergraduates is important enough to be used; then it should ensure that you have been instructed in assessment. How many marks are apportioned for performance in the laboratory, rather than for the write-up? How do you assess technical skills? It is entirely reasonable for you to ask for detailed guidelines from the class organizer.

Other aspects of assessment are equally important. You have to know how to react to a bozo who turns up late to every class, or the trainee lush who is incapable after a lunchtime drinking session (have them evicted because they are a safety hazard). What about the student who works with diligence and care but, because of technical difficulties or maybe incompetence of their partner, fails to acquire any valuable data? Do you give them a set of simulated data to analyse? What do you do about persistent absentees? In most instances, 'pass the buck' is the correct answer; the course organizer is the person who should be making decisions about these individuals. Nonetheless, it falls to you to spot them in the first place.

Taking responsibility

Graduate students must not be seen as supernumerary teachers to lighten the demonstrating and tutorial load of academic staff. The teaching environments in which you will find yourself may be difficult, and need a good deal of planning and preparation. Hopefully, you will be given the detailed guidance that you will need. If not, it could be argued that your department is falling down in its responsibilities to you, and you might consider asking for such instruction. This need not be adversarial, nor does the advice need to be protracted. As a matter of professionalism, you should strive to be as good as you can be in your teaching, as well as your other activities.

There is one final way in which you may take responsibility. Some people are just not cut out to be teachers, and find demonstrating or tutoring a very stressful experience. It may be that you are one of these individuals, and if you feel that you would rather not take on undergraduate teaching, discuss this with your supervisor to see if your wishes can be accommodated. You will, however, lose a valuable opportunity to learn how to communicate, and you should weigh the advantages and disadvantages carefully before deciding to opt out of teaching. As in many aspects of your graduate student training, you may decide to shun the easy way out, because you want to improve the range and depth of your skills.

Moving on

Making the right decision

As your research programme enters its final year, you should be considering your next career move. Assuming that you have decided to remain in science, there are a number of questions that you can pose:

— do I want to continue as a research scientist?

— do I want the security of a permanent position?

— do I want to remain in academic research?

— do I want to specialize in a specific research area?

— do I want to work in another country?

Now is the time for a realistic discussion with your supervisor in which you should outline your career plans, and ask for a frank and honest response. This response may not be what you want to hear, but now is also the time to be really honest with yourself. On average, people change employment about five times in their lives, so you are not necessarily committing yourself to a lifetime career at this stage. However, a change in career direction does seem to get harder as you become more experienced (and older!), and so this is a good opportunity for some hard thinking.

There is a good possibility that you embarked on your postgraduate programme without too much effort. When you originally decided that you wanted to conduct research for a higher degree, you may have visited a few laboratories, but may not have considered applying for any other type of job. If you are going to move out of the relaxed, rather informal environment of academia, be prepared for a shock — the rest of the world works rather differently! If you have never used the services of your institution's Careers Advisors before, this might be a good time to request an interview, and to establish what services are available. The chances are good that you will be impressed by the services placed at your disposal, both in terms of the opportunity for self-appraisal and for the practicalities of job-hunting. Where the Careers Service may fall down is in its knowledge of specialist fellowships and jobs — for this you will have to do your own groundwork.

The answers to the questions listed above will influence the type of employment you will seek; the more restrictive you are the less choice you will have. In general, you will become aware of suitable positions through advertisements in journals such as *Nature* and *Science*. Since you are likely to be reading these journals every week, it does no harm to cast an eye over the classified advertisements to discern the scope and frequency of posts that would meet your criteria.

Most career 'fairs' tend to be focused on undergraduate needs, and it is less likely that you will find a route to a specialized post at such events. The larger scientific conferences often maintain a notice board for advertisement of positions. There is a good chance that the person advertising the post will be present at the conference, which allows for an informal meeting.

The approach outlined above is reactive, inasmuch as you apply for positions as they are advertised. You might want to be less passive about your career development and take greater responsibility for your next move. If you are particularly interested in a research group or a company you might consider making a direct, unsolicited application, enclosing your *curriculum vitae* and a covering letter that explains, in some detail, your reasons for making such a specific application.

Finding your own funding

Postdoctoral fellowships that are advertised in journals will usually be funded by grants that have been awarded for a specific programme of research. As such, there will only be limited opportunities for you to pursue your own interests. Alternatively, you may wish to take greater responsibility for your career, and apply for a grant yourself to conduct research at a laboratory of your choice. This funding will usually be in the form of a postdoctoral fellowship, tenable in an academic or government-funded institution. Sources of such fellowships are charities, the Research Councils and their overseas equivalents such as the National Institutes of Health. Transnational bodies, such as the European Molecular Biology Organization or the Human Frontiers Programme, may offer suitable fellowships. The European Community offers programmes and fellowships that have, as one goal, the encouragement of mobility of research staff. Competition for such prestigious fellowships is intense and it is essential that you discover as much as possible about the 'rules of engagement'. If you have to identify a laboratory to work in, make sure that it meets your criteria and the criteria of the awarding body, that it is prestigious and that you have discussed your intentions with the research group leader. If you are unclear about specific points, write to, fax or telephone someone at the funding organization for clarification. Do not guess or make assumptions that might subsequently mean that you are eliminated from consideration at an early stage for bureaucratic reasons.

It can be difficult to find out about the various schemes that are available. Ask around your department, scrutinize the classified advertisements in journals and check notice boards for specific flyers that might have been sent to your institution. Also, let your supervisor, Head of Department and departmental research correspondent know about your intentions and ask them to let you know about any schemes that are brought to their attention. Try to make contact with someone who holds or has held a postdoctoral fellowship, and seek their advice.

For this type of fellowship, you will usually have to nominate the laboratory in which you would like to work. Because there will be a good deal of correspondence between you and the head of this laboratory, it is essential to make contact as early as you can. Send a letter of introduction outlining your reasons for wanting to work in that laboratory, your *curriculum vitae* and copies of any research papers.

Preparing your *curriculum vitae*

The *curriculum vitae* (c.v.) is normally used as the initial sift of applicants for a job, although many companies will also request that you complete an application form. The appearance and content of your c.v. is therefore crucial if you want to be called for interview. Never send a c.v. without a personal covering letter introducing yourself to the person offering the position. Your application will be sent unchanged to that person, even if it has to be sent initially to a personnel officer.

Undoubtedly, you will already have a c.v. of some description, but this might be a good time to subject it to a critical review. Put yourself in the position of a potential employer, and look at your c.v. with as jaundiced an eye as you can manage. Invite others, whose opinion you respect, to do the same.

A c.v. makes two types of impact: the initial one, when overall appearance is more important than content, and subsequently, the opinion that is formed when the information is read. Aim for clarity, both in content and organization. If you 'lost' a year somewhere, it is better to draw attention to that period than to hope that it will be overlooked — employers have an uncanny knack of spotting these anomalies, and you will have labelled yourself as devious.

Your c.v. should be typed, or still better, prepared on a word-processor so that it can be updated regularly. Ensure that all spelling/typographical errors are corrected, and that you have made effective, but not excessive use of different fonts and styles. Your c.v. must be grammatically correct. In general, there should be no need to use fonts greater than 12pt; indeed, 10pt can look more effective. A smaller font, coupled with effective use of white space can look very attractive. Do not print your c.v. on a low-resolution dot-matrix printer — find a high-resolution laser or inkjet printer. When you photocopy your c.v., take care to ensure that you generate clean white copies, free from smudges — it might be preferable to have copies made at a printing agency. Although of less importance, print/copy your c.v. on to good quality, medium weight paper — not photocopier paper. Have someone, (preferably literate!) read over your c.v. and give you their reactions and opinions.

Prepare your c.v. with a cover page, containing '*Curriculum vitae*', your name and the date (nearest month) in a large and attractive font. It is a good idea to align this text to the right margin — it looks more professional than centre justification and is easier to spot in a sheaf of papers on a desk.

Use subsequent pages liberally. There is little to be gained by compacting the material just to save a page. If each page is only partly filled, updates and embellishments can subsequently be made much more easily.

The contents of your c.v. will change as your career develops, and this may be the time to omit all but the basic details of your school education. Your degree course will still be of interest, as of course will your final degree result. If you undertook an Honours year project, or your course included a period of work experience, describe them in sufficient detail to interest the reader and invite discussion.

Why not offer to give a short seminar, based on your final-year talk?

Details of your current research project are essential. Remember that your postgraduate research programme is primarily a training in research methods and techniques, and emphasize those aspects. Describe the skills that you have learned. In a separate section, outline your research project — the reasoning behind the project, the main discoveries and the implications of those discoveries. Write for a

---

R J Beynon

Curriculum vitae

Dec 1993

scientifically literate individual, but remember that few readers will have your detailed knowledge of this research area.

If you already have publications, include them in a bibliography. Even if these are only poster presentations at a meeting, they indicate that your work is good enough to be offered to your peers, that you have had experience in this form of presentation and that you have attended scientific conferences.

Referees

Your referees must include your supervisor (or, if you are jointly supervised, all of them). It hardly needs to be said that you should have asked them for permission to use their names — this is a formality, but is nonetheless appreciated.

At the interview

If your c.v. and references have worked, you will be called for interview. You should approach this exercise with the same degree of professionalism that you have applied to all aspects of your postgraduate studies. The interview can range from an informal discussion with a research group leader to a two-day exercise in which you will be asked to sit through tests, take part in team activities and meet with staff in the Personnel Department, as well as those with whom you might ultimately be working. In all instances, you should be suitably prepared. You will not impress by asking questions that might easily have been discovered in the company literature that was sent to you. If you are being interviewed for a job in a research laboratory, it will pay to read over some of the recent papers published by the group. The advertisement will always mention something about the subject matter; it would be wise to check some key terms to ascertain the relevance and scope of the subject. Certainly, you will gain credit for having taken the trouble to find out something about the project, and about the person who is advertising the position. Someone in your department may know the company/laboratory to which you have applied — use them to gain inside information.

Most interviews are preceded by a tour around the department and the laboratory. This is an opportunity for you to show your interest in science, and to assess the level of facilities in the department/laboratory, the number of people in the laboratory, the level of activity and the overall work environment. Ask yourself if the research output, determined by a scan of the group leader's name in the literature databases, is commensurate with the number of people in the laboratory. Bear in mind, of course, that published work may reflect past levels of activity; however, a survey of the author list of the papers may be of use in this instance.

At the interview, you can expect to be asked something about your research project. Anticipate this, and prepare a short description of your work, highlighting the important findings. Try to sound enthusiastic about your current work, even if it has lost some of its initial attraction, or if your research programme has not taken the direction that was initially planned. Think about the wider implications of your work, and put it into a broader context. If your research programme was an industrial collaboration, and particularly if you are attending interviews with another company, establish in advance whether there are any aspects of your work that your supervisor(s) might prefer you to avoid discussing.

The interviewer may seem to ask questions that are probing or even adversarial in tone or content. Keep in mind that this is not a personal attack on you; consider the question, and produce a considered reply. Your reaction to questions such as these can make the difference between a job offer or a rejection.

At the conclusion of the interview, you should not expect to be given a decision immediately, but it is reasonable for you and your interviewer to establish a date by when you can expect to hear of the outcome. If you have heard nothing by this date you should make enquiries — letters do get lost in the post. If you have decided, at the end of the visit, that you do not want to be considered for the post, it is only fair that you should write immediately, informing the interviewer of your decision. Do not wait for an offer as a way to assess how well you performed at interview. On the other hand, if you are still interested, write and thank the interviewer for giving you an opportunity to present yourself, reaffirming your interest in the position, and confirming the date when you will be available for employment.

Whether or not you are offered the position, and whether or not you still wish to be considered as an applicant, it is generally accepted that the interviewer will meet your travel expenses for a visit.

Before you move on

One final request. Eventually the time will come for you to leave your laboratory. At this time it is understandable that the focus of your attention will be on the future, but spare a few minutes thought for the people who will continue to work in your laboratory. Throughout your research programme you will have deposited samples and other materials all over the laboratory and department, some perhaps hazardous, some using up resources such as X-ray film cassettes. This is the time to gather together all of this material and discuss with your supervisor exactly what can be disposed of, and what must be retained. It may be some time before anyone returns to these materials, so it is helpful if they are stored in as a stable a form as can be accomplished (frozen cells, DNA under ethanol, and so on). All of these materials should be labelled with your initials and those of your supervisor, your laboratory number and the date. If you have to store a lot of materials in one place, such as a room at –20°C, it is preferable to store everything in a single, clearly labelled box. When the room is cleared out subsequently, there is much less chance of your material being lost, discarded or accidentally defrosted.

As the only person who knows the details of your samples, it is your responsibility to ensure that all hazardous materials are disposed of in the appropriate manner. The same applies to the hazardous chemicals that were purchased for your use — find out whether they will be needed in the future, or whether you should dispose of them before you leave. If there are data sheets for such hazardous chemicals, leave them in the laboratory.

Some tension can develop over the ownership of your laboratory notebooks — do they belong to you or your supervisor? Be realistic about this. Many postgraduate students, and indeed their supervisors, probably never look at these notebooks again. But, if you have papers to write, or if you are continuing in a field where your notebooks would be valuable, you might want to take them with you. If you expect your supervisor to write the papers, then you must leave them in the laboratory. Under no circumstances throw them away; they are a record of your work and it is hard to know whether some of that work, even if never published, might become relevant some time in the future.

Looking back

So, your period of postgraduate study is over, your thesis is complete, papers are written and the degree awarded. Take a moment to reflect on this phase of your life, and try to remember the person you were when you came to the laboratory. Realize just how many skills and new abilities you have acquired, largely through your own efforts. Recognize the role of your supervisor in this transformation.

You may find yourself working in a laboratory in which new graduate students are just starting out. Recall all the lessons that you learned the hard way, and try to help these new students to avoid making the same mistakes. But don't give them answers — offer advice that allows them to analyse the problem and provide their own solutions. This is the way you developed — do not deny them the same experience!

I Index